Allibai Mohanan Vinu Mohan
Wearable Energy Storage Devices

Also of interest

Polymer-Based Solid State Batteries
Daniel Brandell, Jonas Mindemark, Guiomar Hernández, 2021
ISBN 978-1-5015-2113-3, e-ISBN (PDF) 978-1-5015-2114-0,
e-ISBN (EPUB) 978-1-5015-1490-6

Electrochemical Energy Storage.
Physics and Chemistry of Batteries
Reinhart Job, 2020
ISBN 978-3-11-048437-3, e-ISBN (PDF) 978-3-11-048442-7,
e-ISBN (EPUB) 978-3-11-048454-0

Hydrogen Storage.
Based on Hydrogenation and Dehydrogenation Reactions of Small
Molecules
Thomas Zell, Robert Langer (Eds.), 2019
ISBN 978-3-11-053460-3, e-ISBN (PDF) 978-3-11-053642-3,
e-ISBN (EPUB) 978-3-11-053465-8

Electrochemical Energy Systems.
Foundations, Energy Storage and Conversion
Artur Braun, 2019
ISBN 978-3-11-056182-1, e-ISBN (PDF) 978-3-11-056183-8,
e-ISBN (EPUB) 978-3-11-056195-1

Electrochemistry.
A Guide for Newcomers
Helmut Baumgärtel, 2019
ISBN 978-3-11-044340-0, e-ISBN (PDF) 978-3-11-043739-3,
e-ISBN (EPUB) 978-3-11-043554-2

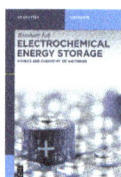

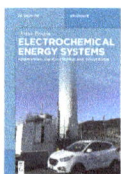

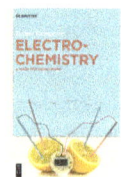

Allibai Mohanan Vinu Mohan

Wearable Energy Storage Devices

—

DE GRUYTER

Author
Dr. Allibai Mohanan Vinu Mohan
EEC, CSIR Central
Electrochemical Research Institute
Karaikudi
Tamil Nadu 630003
India
vinumohan756@gmail.com

ISBN 978-1-5015-2127-0
e-ISBN (PDF) 978-1-5015-2128-7
e-ISBN (EPUB) 978-1-5015-1492-0

Library of Congress Control Number: 2021944793

Bibliographic information published by the Deutsche Nationalbibliothek
The Deutsche Nationalbibliothek lists this publication in the Deutsche Nationalbibliografie;
detailed bibliographic data are available on the Internet at http://dnb.dnb.de.

© 2021 Walter de Gruyter GmbH, Berlin/Boston
Cover image: Sitthiphong / iStock / Getty Images Plus
Typesetting: Integra Software Services Pvt. Ltd.
Printing and binding: CPI books GmbH, Leck

www.degruyter.com

Contents

Abbreviations

0D	Zero dimensional
PET	Polyethylene terephthalate
PEN	Polyethylene naphthalate
PDMS	Polydimethylsiloxane
PS	Polystyrene
SIS	Polystyrene-block-polyisoprene-block-polystyrene
SBS	Polystyrene-block-polybutadiene-block-polystyrene
SIBS	Polystyrene-block-polyisobutylene-block-polystyrene
SEBS	Polystyrene-block-poly(ethylenebutylene)-block-polystyrene
EDLC	Electric double layer capacitors
CNTs	Carbon nanotubes
PEEDOT:PSS	Poly(3,4-ethylene dioxythiophene)-poly(styrenesulfonate)
PANI	Polyaniline
SWCNTs	Single-walled carbon nanotubes
MWCNTs	Multi-walled carbon nanotubes
PAA	Poly(amic acid)
GO	Graphene oxide
PVA	Polyvinyl alcohol
SEM	Scanning electron microscopy
Ppy	Polypyrrole
LIBs	Lithium-ion batteries
PEO	Poly(ethylene oxide)
PVP	Polyvinylpyrrolidone
T_g	Transition temperature
PVDF	Poly(vinylidene fluoride)
PMMA	Poly(methyl methacrylate)
PAM	Polyacrylamide
PU	Polyurethane
ITO	Indium tin oxide
DSSC	Dye-sensitized solar cell
PENGs	Piezoelectric nanogenerators
TENGs	Triboelectric nanogenerators
TEGs	Thermo electric energy generators
PNGs	Pyroelectric nanogenerators
[P(VDF-TrFE)]	Poly(vinylidenefluoride-co-trifluoroethylene)
AC	Alternating current
DC	Direct current

https://doi.org/10.1515/9781501521287-203

1 An overview of wearable energy storage devices

1.1 Introduction

The advent of internet of things is accompanied by tremendous developments in the field of flexible electronics (1–5), particularly, biointegrated wearable devices (6), implantable biomedical devices (7), electronic skin (8), roll-up displays (9), and intelligent mobile devices (10–12). These innovative flexible devices could be effortlessly merged with our routine life, and are useful for diverse advanced applications like personalized healthcare monitoring and therapy (13), auto responsive drug delivery (14), fitness oriented activity monitoring and feedback analysis (15), motion monitoring (16), and interactive rehabilitation (17). The medical diagnostic tools are being focused on noninvasive monitoring of healthcare markers near the patient site, in the home environment (18). Such biointegrated self-controlled devices are suitable for routinely monitoring the physiological condition, early detection, and ensuring timely medical attention (19, 20). The advent of smart wearable technologies facilitates mapping of the complete human health via integrative medicine (14, 21, 22). The wearable physical sensors monitor various physiological parameters such as heart rate (15, 23), blood pressure (24), respiratory rate (25), oxygen saturation (26), temperature (27), and accidental falling (28). The sensors can be controlled by an integrated miniaturized microprocessor that wirelessly transmits the collected quantitative/qualitative health metrics to the smartphone or other wearable gadgets (29).

The wearable chemical sensors monitor the health status by analyzing the bodily fluids such as sweat (30, 31), saliva (32), urine (33, 34), and tears (35, 36). These wearable sensors provide noninvasive tracking of biochemical changes in the human body during physical activities and avoid the painful blood sampling requirements (37). The wealth of information about the biomarker levels and its correlation with the physiological dynamics of body can be utilized for potential point-of-care applications (38, 39). Sweat is a biofluid secreted by the endocrine glands as part of thermoregulation of human body (31). Sweat analysis is mainly associated with fitness monitoring, especially for tracking the electrolyte levels during intense exercise activities (40, 41). In addition it has the potential for clinical monitoring of diseases like cystic fibrosis, via the noninvasive sweat sampling techniques such as iontophoresis (42) and reverse iontophoresis (43). Recent developments include monitoring of drug dosage (44–47), drug abuse (48, 49), depression monitoring (50), infectious disease (51, 52), and many more. Majority of these wearable gadgets are related to the real-time continuous tracking of physiological status, and require stable energy source for their robust operation.

The continuous usage of wearable devices for diverse applications, including data collection and transmission facilities, requires abundant energy over an extended period of time (35, 53). Hence, it is necessary to integrate the wearable electronics

https://doi.org/10.1515/9781501521287-001

with energy storage devices having high energy densities (54, 55). In the last few years, the electrochemical energy storage systems have made tremendous progress, particularly, lithium-ion batteries (LIBs) and supercapacitors (56). The major applications of these devices range from low-power implantable devices to portable apparatus to electrical vehicles to grid-scale power management systems (57). The major developments in the existing energy storage devices are focused on the rigid and bulky prototypes, and their assimilation with the bendable gadgets is challenging. The merging of rigid and bulky energy storage devices along with the bendable electronics is cumbersome and that limits the facile integration with the skin or textile substrates. The forcible assimilation of rigid devices into the non-planar surfaces causes device damage and low performance. Thus, the next-generation energy storage devices are targeted to realize shape conformable and mechanically resilient so as to withstand the external stress that arises from the irregular body movements during normal or exercise activities. An ideal flexible electronics exhibit extraordinary bendability, stretchability, foldability, twistability, and should possess stable performances during mechanical deformations. Ideally, the mechanical properties of the devices need to be similar to that of the wearable substrates. The following sections describe the mechanical properties of the bendable and stretchable materials and other structural configurations to yield stretchability. The mechanics of various structures and their relation with the physical nature and properties of the device components are detailed.

1.2 Flexible and stretchable energy storage devices

Batteries are one of the most suitable candidates for powering flexible electronic devices owing to their high energy and power density, and cycle life. In comparison to batteries, supercapacitors possess high power density and moderate energy density. As the wearable devices mandate flexible and stretchable power sources, it is important to develop fully flexible batteries and supercapacitors with all soft and stretchable constituents. The major components of these devices include current collectors, active electrodes, electrolytes, separator and the outer laminating/package materials. The mechanical properties of all these device components should be similar and any mismatch causes detaching of active materials from the current collector, gel detachment, poor electrical conductivity, and ionic conductivity. The increase in internal resistance of the system results in large ohmic drop and abysmal energy storage activities. In addition, the delaminated electroactive components may penetrate through the separator and which can increase the risk of electrical short circuit. Hence, ideal wearable energy storage devices should consist of equally deformable soft constituents, and must withstand severe mechanical distortions and maintain the charge-discharge activities.

1.2.1 Flexible materials and methods

Flexibility is a property related to rigidity that highlights the deformability of materials. The flexibility of a material is directly related to its mechanical strength, and the stress–strain relationship can be linear elastic, anelastic, or plastic. The initial stage of the development of flexible electronic devices faced several challenges to maintain its flexibility and stable performances especially in terms of electrical conductivity when subjected to multiaxial mechanical deformations. The electrically conducting active materials like zero-dimensional (0D) nanospheres or nanoparticles, 1D nanotubes or nanowires, 2D nanosheets, nanoflakes, and conducting polymers have been utilized for realizing flexible electrochemical devices. Depositing of these nanostructured materials on ultrathin plastic or fabric substrates are broadly studied for various wearable applications. Bendable plastic films like polyimide, polyethylene terephthalate, and polyethylene naphthalate are excellent wearable substrates, owing to their good mechanical integrity under repeated bending cycles, dielectric and insulative properties, ink compatibility, and thermal strength (58). Various thin film coating processes can be exploited for e-skin fabrication such as chemical and physical vapor deposition, sputter coating, or high-precision advanced lithographic techniques.

The textile substrate offers a facile platform for wearable energy storage devices due to its skin conformable nature, feasibility to establish maximum epidermal contact and conformity. There are different types of fabric substrates based on their sources such as wool (animal), cotton (plant), and polyester or nylon (synthetic). For most of the wearable applications, the textile platforms need to isolate with a waterproof layer to avoid skin/sweat contact, parallel electrochemical reactions, interference issues, and electrical short circuit. The flexibility of these plastic or textile substrates mainly depends upon the thickness of the conducting film. Although these platforms are flexible in nature, the mismatching mechanical properties to that of the skin surface limit several epidermal applications. The use of intrinsically stretchable soft polymers can address this problem as they provide secondary skin-like mechanical properties. These synthetic polymers should be biocompatible in nature so as to eliminate the possibility of adverse hazardous or allergic effects.

1.2.2 Intrinsically stretchable materials

The elastomeric polymers having shape memory properties and their composites with 1D or 2D nanomaterials represent key developments in flexible electronics. The intrinsically stretchable materials can withstand enormous strain when subjected to rugged external stress. The viscoelastic properties of the intrinsically stretchable polymers are ideal for binding with the active fillers, and to sustain its integral flexible properties under deformation. Such materials possess weak intermolecular forces, low

Young's modulus, elevated failure strain, and the ability to realign their long chains by themselves to allocate the external stress. The covalent cross linkages enables to reset the primary configurations and avoids the permanent deformation when the stress is taken out. The physical elastomers and the cròss-linked elastomers are potential materials for flexible electronics in which the mechanical strength and the elastic modulus of the former are higher than that of the cross-linked films.

Silicone rubber is a class of organosilicon polymers having wide range of tunable mechanical properties. Polydimethylsiloxane (PDMS) is the most recognized stretchable silicone polymer with high degree of mechanical strength, and is chemically inert, optically transparent, biocompatible, and non-flammable. PDMS is synthesized by reacting dimethyldichlorosilane and water to form a polymer terminated by dimethylsilanolyl groups which is subsequently functionalized with trimethylsilyl chloride. The polymer can be prepared with different viscosities by varying the chain length, and usually possesses tensile strength of 0.8–7.0 MPa, Young's modulus of 0.055–1.9 MPa, and shore hardness of 10–80A when subjected to a strain of 100%.

The thermoplastic homopolymers and block copolymers exhibit three major mechanical properties. The polymers can be stretched to moderate elongations and upon removal of external stress they return close to the original shape. The polymers are processable when subjected to elevated temperatures but remain chemically stable during processing. Also, the exceptional mechanical resiliency facilitates sustaining the physical characteristics, even though exposed to longtime abnormal mechanical stress below the yield strength of the polymer. The thermoplastic polymers are a class of copolymers having characteristics of a plastic and a rubber, resulting in both thermoplastic and elastomeric in nature. These versatile properties of thermoplastics allow constructing different flexible shapes and forms, including thin films, by exploiting wide varieties of techniques like injection molding, mold casting, compression molding, screen-printing, spin coating, and extrusion molding. Polyurethane is the first class of thermoplastic elastomers and then styrene block copolymers are emerged. The urethane group is the major repeating unit in polyurethane elastomers and is prepared by reacting isocyanate resins with polyols (59). The elastomeric polyurethane can be tuned into different engineered products owing to their impeccable physical properties. The malleable nature of polyurethane provides good processability, including extrusion and injection molding, screen-printing, and inkjet printing.

The block copolymers of styrene are recently utilized much for manufacturing stretchable electronic devices for sensing and energy related applications. The copolymers of styrene and butadiene show a two-phase microstructure due to the incompatibility between the individual units. The low polystyrene content is preferable, and polybutadiene predominates in its physical properties. Triblock copolymers having two short rigid polymer blocks end-capped with a long soft central backbone are another predominant elastomeric matrix, and the elasticity is resulted by the presence of hydrophobic polystyrene units within the soft long chain block components. The important triblock elastomers are polystyrene-block-polyisoprene-

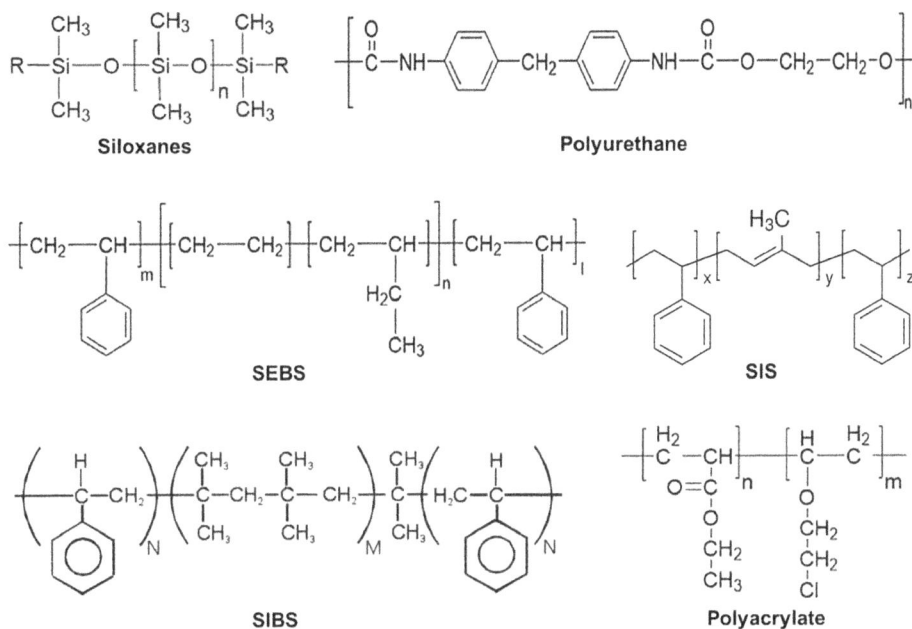

Figure 1.1: Chemical structures of stretchable polymers (reproduced with permission from ref. (60, 61)).

block-polystyrene (SIS), polystyrene-block-polybutadiene-block-polystyrene (SBS) polymers, polystyrene-block-polyisobutylene-block-polystyrene (SIBS), and poly-styrene-block-poly(ethylenebutylene)-block-polystyrene (SEBS). The appreciable solubility of these polymers in several organic solvents provides effortless processabil-ity to configure them into ultrathin films by coating or printing methods. Polyacrylic elastomers are also exhibit elastic behavior, and generally prepared via emulsion or suspension polymerization methods. Polyacrylates show resistance toward both heat and oils with a high endurance temperature of ≈175 °C. The polyacrylates possess typ-ical modulus in the range of 0.6–3.6 MPa at a 100% strain and the elongation at break can be tuned from 150% to 440%. Figure 1.1 shows the chemical structures of major elastic polymers (60, 61), and Table 1.1 summarizes the mechanical prop-erties of synthetic elastomeric substrates (60, 62–71).

1.2.2.1 Stretchable conducting thin films

The electrical conductivity and the mechanical properties of the electrodes are essential for maximizing the performances of energy storage devices. The use of stretchable elec-trodes with low contact resistance and good electrical stability over a wide potential window can minimize the strain-induced device failure. The nature of the con-ducting species and the stretch enduring binders, and other properties like adhe-sion, surface roughness, film thickness, and film patternability, should be considered

Table 1.1: Mechanical properties of different elastomeric polymers.

Polymer material	Brand name	Modulus at 100% strain	Elongation at break [%]	Tensile strength [MPa]	Shore hardness	References
Silicone rubber	Ecoflex	0.055–0.1	800–1,000	0.8–2.4	00–10–5A	(62)
	Sylgard 184 (PDMS)	0.6–3.6 (1.84)	80–170 (120)	3.5–7.7 (7.1)	44–54A (44A)	(63)
	Smooth-sil	1.2–1.9	300–500	4.5–5.0	35–60A	(60)
	Mold Max	0.24–1.31	250–529	3.3–4.0	10–40A	(60)
	Dragon Skin	0.15–0.6	364–1,000	3.3–3.8	10–30A	(64)
	Elastosil	–	1.5–7.0	50–600	20–80A	(65)
PU	Plei-Tech	1.7–13.8	400–720	18.6–51.0	60–97A	(60)
SBS	Kraton D	1.2–2.9 (300%)	600–880	2–33	70–75A	(66)
SIBS	Kraton D	0.8–2.0	≈1,200	3–11	29–46A	(67)
SEBS	Kraton G	2.4–5.6 (300%)	500–750	10.3–35	35–69A	(68)
PVDF–HFP	Kynar, Solef	–	200–800	14–48	50–70A	(69)
Polyacrylate copolymer	Vamac	2.7–7.2	208–440	9.8–18.2	59–79A	(70)
Polyacrylate	HyTemp	≈3.6	150–250	7–14	50–80A	(71)

during the designing and fabrication stages. Conductive nanoscale materials like metal nanowires, metal nanoparticles, carbonaceous nanomaterials, conductive polymers, and the composite of different materials are generally utilized for realizing conducting thin films. In order to withstand the mechanical strain, the conductive fillers are dispersed or embedded into a stretchable polymer matrix. The mechanical deformability of the elastomers maintains the electrical conductivity and the device performances. Table 1.2 discloses various conducting hybrid films developed by merging conductive materials and polymer substrates (72–105). These composite films are appropriate for preparing stretchable electrodes or current collectors for wearable energy storage devices.

Table 1.2: The performances and mechanical properties of various elastomeric conductors.

Materials	Resistance	Transmittance	Stretchability	Formable	References
AgNWs	$35 \ \Omega \ cm^{-1}$	80%	20%	Thin film	(74)
AgNWs/PDMS	$7.5 \ \Omega \ cm^{-1}$	80–90%	50%	Thin film	(75)
CuNWs	$220 \ \Omega \ cm^{-1}$	91%	17%	Thin film	(76)
CuNWs/PU	$10 \ 2 \ \Omega \ cm^{-1}$	68.7–84.5%	60%	Thin film	(77)
AgNWs/PDMS	–	Non-transparent	50%	Thin film	(78)
AgNWs/PDMS	$0.24 \ \Omega \ cm^{-1}$	Non-transparent	50%	Thin film	(79)
AgNWs/PDMS	$26.1 \ \Omega \ cm^{-1}$	85.8%	30%	Thin film	(80)
AgNWs/PUU/PDMS	$5–425 \ \Omega \ cm^{-1}$	60–90%	50%	Thin film	(81)
AgNWs/PDMS	$9 \ \Omega \ cm^{-1}$	81–90%	20%	Thin film	(82)
PEDOT:PSS/soft polymer	$75 \ S \ cm^{-1}$	78%	55%	Thin film	(83)
Graphene–AgNWs hybrid	$1 \ \Omega \ cm^{-1}$	91%	80%	Thin film	(84)
Graphene–AgNWs hybrid	$33 \ \Omega \ cm^{-1}$	100%	95%	Thin film	(85)
Ag flakes/fluorine rubber composite	$738 \ S \ cm^{-1}$	Non-transparent	215%	Thin film	(86)
R-GO/nanocellulose	–	Non-transparent	100%	Thick film	(87)
R-GO microtubes/PDMS	–	Non-transparent	50%	Thick film	(88)
Graphene foam/PMDS	$1.5 \ k\Omega$	Non-transparent	70%	Thick film	(89)
PU-PEDOT:PSS/SWCNTs	–	62%	100%	Thin film	(90)
AgNWs/AgNPs/SBS	$2,450 \ S \ cm^{-1}$	Non-transparent	100%	Fiber	(91)
AgNWs/PEDOT:PSS/PU	$36 \ \Omega \ cm^{-1}$	75%	130%	Thin film	(92)
PDMS/AgNWs/PDMS	$7.5–246 \ \Omega$	Non-transparent	70%	Thin film	(93)
CNT/R-GO/porous PDMS	$27 \ S \ m^{-1}$	Non-transparent	50%	Thick film	(94)
AgNP or PtNP/R-GO	$3,012 \ S \ cm^{-1}$	Non-transparent	35%	Thick film	(72)

Table 1.2 (continued)

Materials	Resistance	Transmittance	Stretchability	Formable	References
PEDOT:PSS/PU	50 S cm^{-1}	Non-transparent	200%	Thick film	(95)
Carbon-black/styrene butadiene composite	40 S cm^{-1}	Non-transparent	200%	Thick film	(96)
AgNP/SWCNT/PU composite	620 S cm^{-1}	Non-transparent	90%	Thick film	(97)
PEDOT:PSS/Acrylamide organogels	–	Non-transparent	150%	Thick film	(98)
CNT–PMIA core–shell	109 S cm^{-1}	Non-transparent	150%	Fiber	(99)
PEDOT:PSS blended Triton X-100	16 Ω cm^{-1}	Non-transparent	50%	Thin film	(100)
Ni electroplated onto porous PDMS surface	–	Non-transparent	80%	Thin film	(101)
Ag/PDMS composite	2 Ω cm^{-1}	Non-transparent	25%	Thin film	(102)
Ni/PDMS composite	20 Ω	Non-transparent	100%	Thin film	(103)
Silver salt/polystyrene-block-polyisoprene-block-polystyrene	0.8 Ω cm^{-1}	Non-transparent	200%	Thin film	(104)
Ag flakes/fluorine rubber composite	738 S cm^{-1}	Non-transparent	215%	Thin film	(86)
SWCNTs/fluorinated copolymer composite	100 S cm^{-1}	Non-transparent	100%	Thin film	(105)
SWCNTs/fluorinated copolymer composite	57 S cm^{-1}	Non-transparent	134%	Thin film	(73)

1.2.3 Design-induced stretchability

A deterministic engineering approach is promising for constructing strain resistant structures by introducing design-induced flexibility. This method exploits the concept of structural mechanics associated with micro/nanoscale inorganic films that are embedded on the stretch enduring substrates. The three-dimensional wavy structures provide good flexibility which can be prepared by patterning thin films on the prestained substrate. The release of the strain after microfabrication realizes the wavy configuration (106). The conducting structural components can be prepared in various geometries like membrane, ribbon, or wires, and these shapes can mitigate the external strain via controlled, nonlinear buckling and/or in-plane bending actions (107). Figure 1.2 shows the representative examples of (a) wavy and (b) ribbon-shaped patterns that have been reported for different applications (106). The island-bridge concept also offer design-based stretchability which consists of active island regions, interconnected with linear, serpentine shaped, or coiled interconnects (108). Upon mechanical deformation, the curved bridges withstand major portion of the developed strain and protect the active region from damage and thus maintain its performances (Figure 1.2c) (107). The free-standing nature of the serpentine interconnects is essential for yielding complete advantages of the island-bridge system. Hence, a sacrificial layer (mostly water-soluble polymer) is usually utilized for selective bonding of active islands on the elastomeric substrate.

The same core idea of strain management can be extended in mechanically assembled 3D structures. The 3D helical coils provide advantages over the planar 2D interconnect by qualitatively improving the buckling processes (109). In 2D curved structures, majority of the strain concentrates at the arc regions and their physical bonding to the substrate restricts the buckling process. But the helical coils facilitate uniform distribution of strain and its free-standing nature enables effortless displacements and enhanced stretchability (Figure 1.2d). Mesh designs are another configuration for stress management where the 2D interconnected bridges undergoes in-plane rotations, rather than out-of-plane displacements (110). Park et al. developed a Pt nano-mesh that can be used for stretchable and transparent energy storage devices (Figure 1.2e) (111). Another promising approach for realizing conductive current collector is the use of liquid alloy in a microchannel which is embedded within the stretchable substrate. Upon stretching, the physical flow of liquid alloy along the deformations maintains the electrical contact and device performances (112). Lee et al. developed a super stretchable triboelectric energy harvesting device by using Galinstan as a liquid metal (Figure 1.2f) (113). A new frontier of wearable energy storage devices is focusing on the development of design-induced stretchable materials and its amalgamation with rigid active constituents (114–117).

1.3 Bending mechanics of energy storage devices

The external bending motion on a monocomponent system causes deformation throughout the entire device and the internal stress is developed to resist the shape deformation. Majority of the stress is localized at the non-uniform regions and the stress is heterogeneously distributed throughout the system. According to

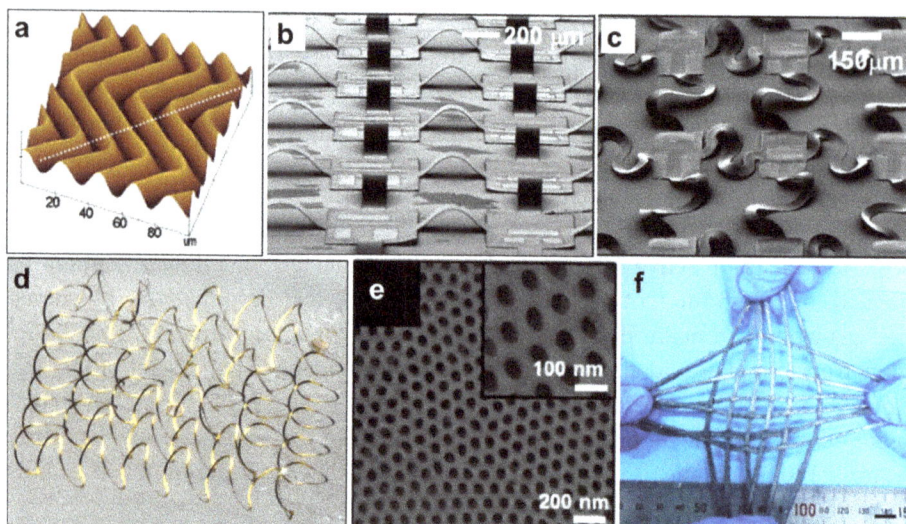

Figure 1.2: (a) AFM image of a 2D wavy Si nanomembrane (reproduced with permission from ref. (106)). SEM images of (b) non-coplanar mesh design integrated with PDMS substrate in an undeformed state and (c) non-coplanar interconnects with serpentine bridge structures (reproduced with permission from ref. (107)). (d) Optical image of lithographically defined multilayer 3D coils of polyimide/Au bonded on a silicone substrate (reproduced with permission from ref. (109)). (e) SEM image of Pt nanomesh structure (reproduced with permission from ref. (111)). (f) Optical image of textile-like mechanical energy harvester at stretched state (reproduced with permission from ref. (113)).

Hooke's law, the elastic modulus of a material relates the ratio between stress and strain within the elastic limit. The energy storage devices usually consist of multiple layers including active layer, electrolyte, and substrate (Figure 1.3a), and their bending mechanics is complex in nature due to the difference in properties like elastic modulus and Poisson's ratio. When a multilayer system is bent to a cylinder of radius R, then the outer surface bears tensile strain and the inner locations endures compressive strain (Figure 1.3b). The device possesses a mechanical neutral plane without any uniaxial strain, and the point of this plane is depending on the thickness of each layers and their Young's modulus (Figure 1.3c). The mismatch in elasticity of different layers results in concentration of stress on the layer or at the interface between the films. When the internal strain increases beyond the tolerance limit, the device structures undergo crack or delamination, and weakening of their electrical properties. The difference in film thickness of active material and the substrate also causes film rupture or interfacial deterioration. There are two modes of failures in multilayer devices such as tensile and compressive strains. Due to the tensile strain, the fragile sites of the constituent layers crack, and the compression strain results in large buckling of unbonded parts. The compression strain becomes proportional to the bonding strength and the strain in the convex surface is equal to the distance b from the neutral plane divided by R, as given by:

$$\varepsilon = \frac{b}{R} \tag{1.1}$$

For practical bending mechanics evaluation, arbitrary conclusions are generally avoided because the mechanical behaviors of devices are related to the defect state, temperature manufacturing process, and environment. Hence, it is important to conduct replicate measurements using multiple samples for precise stress-strain analysis.

The mechanical bending characteristics of the flexible energy storage devices can be expressed by their bending endurance under a particular radius. Major parameters describing the bending status are the end-to-end distance (L) along the bending direction, the bending angle (θ), and the bending radius of curvature (R). These factors of a flexible device are schematically represented in Figure 1.3b as a mechanical beam. The parameter L is widely used to demonstrate the bending condition in any devices having different shapes like sheet, thin film, wire-shaped and fiber-type architectures. Although the bending analysis involves simple apparatus and easy operation, the L value endows limited information about the bending mechanics and inappropriate for widespread measurements of strain and failure modes. For soft and thin devices having non-uniform mechanical deformation, the L value is insufficient to comprehend the accurate bending state.

Bending angle is defined as the rotating angle of the moving end, and the typical measurement process for bending angle is illustrated in Figure 1.3d. The length of the flexible devices significantly affects the mechanical strain at the peak position and influences the determination of bending angle. For instance, the ultra-long architectures endure small bending strain even for a large bending angle because of its large bending radius. Hence, the bending angle also provides a rough estimation of bending status. Bending radius of curvature R is typically related to the bending strain of flexible devices except for the intrinsic young's modulus and thickness of devices. Thus, bending radius is the most suitable parameter for understanding the failure modes of flexible devices and to assess the relationship between structural configuration and mechanical property. A simple and intuitive method to measure the bending radius is winding the devices around cylinders with known diameters (mandrel method). Such bending operation can be executed manually without depending on any complex equipment, and this method can be applied to sheet-like, fiber-shaped, and thin/soft devices. Collapsing radius geometry is another method to determine the radius, which involves fixing of devices between two parallel plates, followed by squeezing with a linear actuator. In this, half of the distance between both the poles gives the bending radius. However, the bending geometry is irregular and not precisely cylindrical, and the stress and deformation at various regions of the devices are non-uniform.

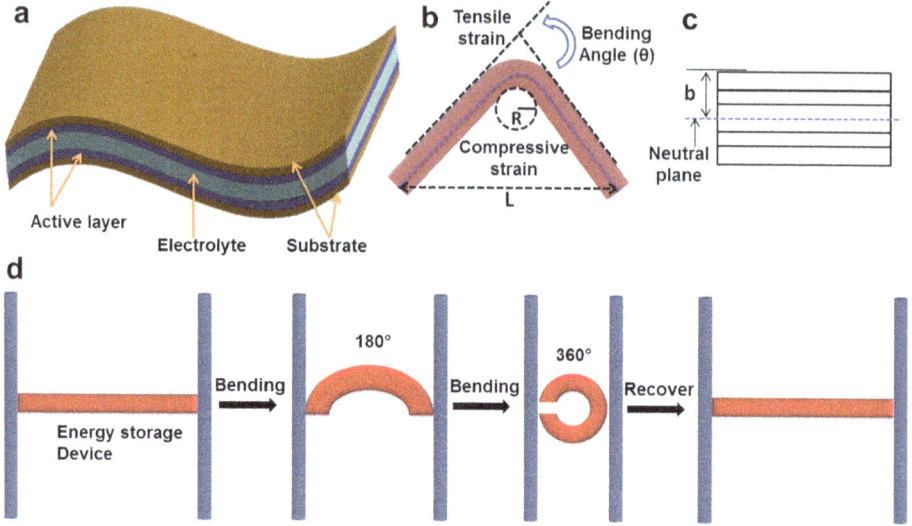

Figure 1.3: (a) Schematic illustrations of (a) multiple layers of flexible energy storage devices and (b) the bending mechanics of flexible system with major parameters (L, θ, and R) to describe the bending state. (c) Represent the multilayer system showing the mechanical neutral plane. (d) Schematically representing different bending states in flexible fiber-shaped energy storage devices.

X–Y–θ method provides more accuracy and uniform bending angle to determine the radius based on an arc-shaped motion with controllable bending speed. During testing, one end of the device is fixed and the moving side is controlled by a rotary motorized actuators. For a typical X–Y–θ configuration, the fixed end is defined as original point, the length between both poles is L_0, and the early horizontal direction can be represented as X axis, and θ is the angle between horizontal axis and the tangent of the moving part. The spatial coordinates are controlled in a (Xi, Y_i, θ_i) pattern of a circumference and can be calculated from the equations (1.2) to (1.4):

$$X_i = \frac{L_0}{\theta_i} \sin\theta_i \tag{1.2}$$

$$Y_i = \frac{L_0}{\theta_i} (1 - \cos\theta_i) \tag{1.3}$$

$$\theta_i = \frac{L_0}{R} \tag{1.4}$$

In X–Y–θ method, the bending angle is restricted between 0° and 180° to circumvent the contact of sample sites, and the smallest bending radius should be less than L_0/π. This technique is extensively used to distinguish the bending property of flexible display devices and can also be extended to flexible energy storage devices with uniform mechanical deformation.

1.4 Mechanics of stretchable energy storage devices

The key challenge associated with the stretchable electronics is the mismatch between the soft components and intrinsic hard inorganic materials with a fracture strain of 1%. Numerous mechanics strategies have been introduced to construct stretchable inorganic electronic devices. The approaches can be primarily classified into two categories such as wavy design and island-bridge design. In wavy configuration, the active rigid materials are transfer-printed to a pre-stretched elastomeric platform, and upon releasing the elastomer forms a wavy arrangement that can withstand enormous strain and mechanical deformations via changes in wavelength and amplitudes. The fabrication of stretchable wavy ribbons is demonstrated in Figure 1.4a. The ribbon is perfectly bonded on a pre-strained substrate, and the shrinking of the substrate causes compression of ribbons to form a wavy configuration through a non-linear out-of-plane buckling process.

1.4.1 Mechanics of small distortion

In case of small deformations of wavy ribbon, the thin film thickness is much smaller when compared to other characteristic dimensions like wavelength, and can be modeled as an elastic non-linear von Karman beam. The substrate can be modeled as a semi-infinite bulky solid since its thickness (mm scale) is greatly larger than that of the coated film (μm). The total energy associated with the configuration includes bending energy (U_b) and membrane energy (U_m) in the film and strain energy (U_s) in the substrate. The out-of-plane displacement of the wavy thin film, upon releasing the pre-strain, can be represented as equation (1.5), whereas x is the coordinate along the film direction, A represents the amplitude, k is the wavelength, and $k = 2\pi/\lambda$ is the wave number.

$$w = A\cos(kx) = A\cos\left(\frac{2\pi x}{\lambda}\right) \tag{1.5}$$

The bending energy U_b can be represented by equation (1.6), for a thin-film thickness of h_f and Young's modulus and Poisson's ratio of E_f and m_f, respectively. The pre-strained compliant substrate holds pre-strain of ε_{pre}, Poisson's ratio of m_s, and modulus of E_s.

$$U_b = L_0 \frac{1}{\lambda} \int\limits_0^\lambda \frac{\bar{E}_f h_f^3}{24} \left(\frac{d^2 w}{dx^2}\right)^2 dx = \frac{\pi^4 \bar{E}_f h_f^3 A^2}{3\lambda^4} L_0 \tag{1.6}$$

where L_0 is the film length and $\bar{E}_f = E_f/(1 - V_f^2)$ represents the plane-strain modulus of the film. The membrane energy U_m in the film can be determined by equation (1.7), where ε_m is the membrane strain and N_m is the membrane force.

$$U_m = L_0 \frac{1}{\lambda} \int_0^\lambda \frac{1}{2} N_m \varepsilon_m dx = \frac{1}{2} \bar{E}_f h_f \left(\frac{\pi^2 A^2}{\lambda^2} - \varepsilon_{pre} \right)^2 L_0 \tag{1.7}$$

The strain energy associated with the substrate can be determined by following equation (1.8):

$$U_s = \frac{\pi}{4\lambda} \bar{E}_s A^2 L_0 \tag{1.8}$$

where $\bar{E}_s = E_s / (1 - V_s^2)$ is the plane-strain modulus of the flat substrate.

The minimization of the total energy provides the buckle amplitude A and wavelength λ, as

$$\lambda = 2\pi h_f \left(\frac{\bar{E}_f}{3\bar{E}_s} \right)^{1/3}, \quad A = h_f \sqrt{\frac{\varepsilon_{pre}}{\varepsilon_c} - 1} \tag{1.9}$$

The peak strain (ε_{peak}) is the sum of membrane strain and bending strain and can be determined by:

$$\varepsilon_{peak} \approx 2\sqrt{\varepsilon_{pre}\varepsilon_c} \tag{1.10}$$

When the buckled system is subjected to a strain ($\varepsilon_{applied}$), the wavelength and amplitude change to:

$$\lambda = 2\pi h_f \left(\frac{\bar{E}_f}{3\bar{E}_s} \right)^{1/3}, \quad A = h_f \sqrt{\frac{\varepsilon_{pre}}{\varepsilon_c} - 1} \tag{1.11}$$

And the peak strain in the ribbon becomes:

$$\varepsilon_{peak} \approx 2\sqrt{(\varepsilon_{pre} - \varepsilon_{applied})\varepsilon_c} \tag{1.12}$$

1.4.2 Mechanics of large distortion

While large pre-strain is applied, the wavelength shows a linear dependency on the strain which decreases with the increase of the deformations. The strain-dependent wavelength can be accounted for three factors such as finite geometry change like different stress-free or strain-free states for thin film and substrate, non-linear strain–displacement dependency, and non-linear constitutive model for the substrate. The finite geometry alteration dominates among all three aspects, and the out-of-plane dislocation of the buckled thin film can be represented as equation (1.13):

$$w = A\cos\left(\frac{2\pi x}{\lambda}\right) = A\cos\left(\frac{2\pi x'}{(1+\varepsilon_{pre})\lambda}\right) \tag{1.13}$$

where the coordinate x in the strain-free state for the substrate and x in the strain-free state for the ribbon are related by $x' = (1+\varepsilon_{pre})x$. The bending energy and membrane energy in the film can be obtained as equations (1.14) and (1.15), where $(1+\varepsilon_{pre})L_0$ is the original length of the thin film.

$$U_b = \frac{\pi^4\,\bar{E}_f\,h_f^3\,A^2}{3\left[(1+\varepsilon_{pre})\lambda\right]^4}(1+\varepsilon_{pre})L_0 \tag{1.14}$$

$$U_m = \frac{1}{2}\bar{E}_f h_f \left[\frac{\pi^2 A^2}{(1+\varepsilon_{pre})^2\lambda^2} - \frac{\varepsilon_{pre}}{1+\varepsilon_{pre}}\right]^2 (1+\varepsilon_{pre})L_0 \tag{1.15}$$

The wavelength and amplitude can be determined by minimizing the total energy as follows:

$$\lambda = \frac{2\pi h_f}{1+\varepsilon_{pre}}\left(\frac{\bar{E}_f}{3\bar{E}_s}\right)^{1/3}, \quad A = h_f\sqrt{\frac{\varepsilon_{pre}}{(1+\varepsilon_{pre})\varepsilon_c} - 1}, \tag{1.16}$$

The peak strain in the buckled structure changes to:

$$\varepsilon_{peak} \approx 2\sqrt{\frac{\varepsilon_{pre}}{1+\varepsilon_{pre}}\varepsilon_c} \tag{1.17}$$

The out-of-plane displacement when the buckled system is affected by the applied strain $\varepsilon_{applied}$ can be determined as:

$$w = A''\cos\left(\frac{2\pi x''}{y''}\right) = A''\cos\left[\frac{2\pi x''(1+\varepsilon_{applied})}{\lambda''(1+\varepsilon_{pre})}\right] \tag{1.18}$$

where $x'' = (1+\varepsilon_{applied})x$ is the coordinate in the stretched condition. The bending, membrane, and substrate energies can be calculated as follows:

$$U_b = \frac{\pi^4\bar{E}_f h_f^3 A''^2 (1+\varepsilon_{applied})^4}{3\left[(1+\varepsilon_{pre})\lambda''\right]}(1+\varepsilon_{pre})L_0 \tag{1.19}$$

$$U_m = \frac{1}{2}\bar{E}_f h_f \left[\frac{\pi^2 A''^2(1+\varepsilon_{applied})^2}{(1+\varepsilon_{pre})^2\lambda''^2} + \frac{\varepsilon_{applied}-\varepsilon_{pre}}{1+\varepsilon_{pre}}\right]^2 (1+\varepsilon_{pre})L_0 \tag{1.20}$$

$$U_s = \frac{\pi}{4\lambda''}\bar{E}_s\,A''^2(1+\varepsilon_{applied})L_0 \tag{1.21}$$

The minimization of the total energy results in the wavelength and amplitude:

$$\lambda'' = \frac{2\pi h_f \left(1 + \varepsilon_{applied}\right)}{1 + \varepsilon_{pre}} \left(\frac{\bar{E}_f}{3\bar{E}_s}\right)^{1/3}, A = h_f \sqrt{\frac{\varepsilon_{pre} - \varepsilon_{applied}}{\left(1 + \varepsilon_{pre}\right)\varepsilon_c} - 1} \tag{1.22}$$

When the tensile strain increases, the wavelength increases but the amplitude reduces. When the tensile strain reaches the sum of pre-strain and the critical strain, the amplitude becomes zero and further strain $\varepsilon_{fracture}$ leads to fracture the film. Hence, the stretchability is the sum of ε_{pre}, $\varepsilon_{fracture}$, and ε_c. The peak strain in the ribbon can be determined by equation (1.23):

$$\varepsilon_{peak} \approx 2\sqrt{\frac{\varepsilon_{pre} - \varepsilon_{applied}}{1 + \varepsilon_{pre}}\varepsilon_c} \tag{1.23}$$

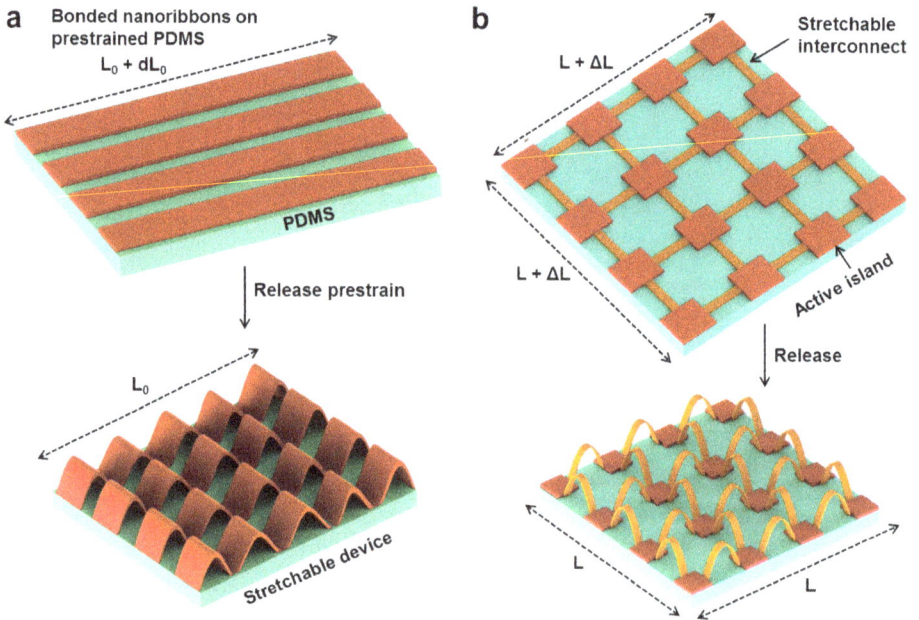

Figure 1.4: Schematic demonstration of (a) fabricating wavy ribbons and (b) island-bridge design on an elastomeric PDMS substrate.

1.4.3 Mechanics of island-bridge pattern

The wavy structured devices provide stretchability along one direction of which the pre-strain is applied. In order to attain the stretchability in all directions the island-bridge design of the system is proposed (118). In this pattern the active areas are firmly bonded with the pre-strained stretchable substrate, called islands, and are interconnected with the loosely or non-bonded bridges. Once the strain is released, the interconnects are buckled and displaced to out of plane of the substrate. These buckled bridges hold maximum of the strain developed during deformations and protect the active islands from deteriorations.

1.4.3.1 Island-bridge design having linear interconnects

Song et al. developed an analytical model for island-bridge pattern having straight interconnects to describe the underlying mechanics of the system (119). The profile of the buckled system for a strain-free bridge X in the pre-stretched configuration and the buckled configuration x when the strain is released can be expressed as equation (1.24):

$$w = \frac{A}{2}\left(1 + A\cos\left(\frac{2\pi x}{L_{bridge}}\right)\right) = \frac{A}{2}\left(1 + A\cos\left(\frac{2\pi x}{L_{bridge}^0}\right)\right) \qquad (1.24)$$

The total energy of the bridge consists of the bending energy and membrane energy, and can be stated as:

$$U_b = \frac{E_{bridge}h_{bridge}^3}{12}\frac{\pi^4 A^2}{\left(L_{bridge}^0\right)^3} \qquad (1.25)$$

$$U_m = \frac{1}{2}E_{bridge}h_{bridge}L_{bridge}^0\left[\frac{\pi^2 A^2}{4\left(L_{bridge}^0\right)^2} + \frac{L_{bridge} - L_{bridge}^0}{L_{bridge}^0}\right]^2 \qquad (1.26)$$

The minimization of total energy in the bridge region provides the amplitude,

$$A = \frac{2L_{bridge}^0}{\pi}\sqrt{\frac{L_{bridge}^0 - L_{bridge}}{L_{bridge}^0} - \frac{\pi^2 h_{bridge}^2}{3\left(L_{bridge}^0\right)^2}} \qquad (1.27)$$

The peak strain in the bridge can be determined by:

$$\varepsilon_{bridge}^{max} \approx 2\pi\frac{h_{bridge}}{L_{bridge}^0}\sqrt{\frac{L_{bridge}^0 - L_{bridge}}{L_{bridge}^0}} \qquad (1.28)$$

When the pre-strain is released, the bridge length is reduced from L_0 to L_{bridge}, which provides the pre-strain $\varepsilon_{pre} = \dfrac{L_{bridge}^0 - L_{bridge}}{L_{bridge}^0}$. Then the maximum strain in the bridge changes to:

$$\varepsilon_{bridge}^{max} \approx 2\pi \frac{h_{bridge}}{L_{bridge}^0} \sqrt{\frac{\varepsilon_{pre}}{1 + \varepsilon_{pre}}} \tag{1.29}$$

1.4.3.2 Lateral buckling of straight interconnects

When the island-bridge system is subjected to shear or diagonal stretch, the bridges endure lateral buckling which is much different from the common Euler buckling that involves simple sinusoidal displacement. The lateral buckling causes out-of-plane bending and high degrees of torsion with complex dislocations. Su et al. developed a systematic method for understanding the post-buckling of beams based on the nonlinear equilibrium equations (120). The perturbation method provides the amplitude of buckled bridges. Chen et al. derived analytical model for initial and post-buckling of interconnects under shear (121). The initial buckling process is analyzed via the equilibrium equations, whereas the post buckling is investigated by minimization of the potential energy up to fourth power of the dislocation.

1.4.3.3 Buckling of serpentine interconnects

The use of serpentine-shaped interconnects instead of straight bridges augments the stretchability of the system. The serpentine bridges possess advantages over the straight pattern: (1) the serpentine bridges are longer than the linear one, and hence they can withstand high degrees of pre-strain; (2) when the external strain becomes equal to the pre-strain of the substrate, the straight bridges turn into flat and the excess strain can damage the device, but the serpentine bridges provide unwinding of the curved regions and thus hold much higher level of stretchability. The complex geometry of the serpentine pattern limits the analytical modeling of the deformation of serpentine bridges. The finite element simulations are widely exploited for understanding the underlying mechanics of buckling of serpentine-shaped interconnects.

1.4.3.4 Buckling of fractal interconnects

The recent development in the island-bridge concept includes the incorporation of fractal interconnects that provides high level stretchability and large areal coverage of islands. The fabrication of interconnects in Hilbert fractal design with multiple levels of iterations provides enormous stretchability. The first level pattern possesses a U shape, and the higher levels correspond to self-similar assemblies of the same patterns. For fixed spacing between device islands, when each successive level rises, the overall length of fractal interconnects enhances approximately by a factor of 2. The unraveling of the fractal interconnects provides accommodates high degrees of strain and increases the system-level stretchability. The buckling of fractal interconnects begins at the highest level and then propagates to lower levels.

This mechanism is completely reversible and the fractal pattern restores upon unloading the external stress.

The fractal designs can be engineered to accommodate elastic strain along a selected dimension and also to biaxial and radial deformation modes. Rogers et al. reported that the choices of topologies can be extended to a broad range, from lines (Koch, Peano, Hilbert) to loops (Moore, Vicsek), and branch-like meshes (Greek cross) (108). The approximate fractal dimensions in such finite-iterative curves range from 1.5 to 2. Hence, this fractal design method can be used for specific electronic applications via integration and interdigitation of various structures.

1.4.4 Analysis of mechanical properties

The mechanical properties of materials are related to the response of that material to an applied force, termed as loading. The most significant mechanical properties are strength, hardness, impact resistance, ductility, and fracture toughness. The precise analysis of these materials' properties is crucial for determining its potential use and expected lifespan (60). The shape of the materials (film, plate, wire, etc.) largely influences their mechanical properties, and the direction, at which the force is exerted, as well as loading rate and temperature is another critical parameters. This necessitates conducting the mechanical stress measurements with identical conditions, and choosing a range of values for a given parameter. Also, the standards and statistical reproducibility of the newly reported materials are needed to be established for optimizing the stretchable device performances. The most significant loading characteristics are bending, tension, compression, shear, and torsion (Figure 1.5a). The bending mechanics is primarily important for flexible electronics, and the other two mostly scrutinized for stretchable electronics. Tension is a type of loading where two ends of an object are pulled away from each other. Compression is the opposite operation of tensile loading where the material is pressed together. The static loading involves applying a constant force to a material, but the cyclic or dynamic loading utilizes the force that fluctuates periodically.

The parameter stress is a physical quantity which expresses the internal forces among the adjacent particles of a continuous material that exert on each other. Stress is used to quantify an applied force to a certain cross-sectional area of an object, and which equals the internal distribution of forces inside the object that reacts to the loads exerted. Hence, the stress distribution depends on the nature of loading and the shape of object. When a wire is subjected to a tensile loading, the tensile stress is uniformly distributed (Figure 1.5b). The stress is determined by stress $= F/A$ with a unit of N m^{-2} or Pascal. But when a bar is loaded in bending deformation, the stress distribution is varied with the distance perpendicular to the normal axis. Strain (ε) is the measure of the deformation of an object to an applied stress. For a wire under tension, strain is the ratio of total elongation and the initial length.

Several important mechanical properties can be determined by a tensile test, such as the elastic or Young's modulus, yield strength, elastic limit, elongation, and tensile strength. The result of a tensile load applied to an object while measuring the elongation is called a stress–strain curve (Figure 1.5c). Each material possesses a characteristic curve, and the linear part indicates an elastic deformation. Once the stress is released, the elastic material recovers its original shape and thus it obeys Hooke's law. The slope of the linear portion is called the Young's modulus (E), which is a measure of stiffness of a material. But the brittle materials show minimal or no plastic deformation, and they fractures near the end point of the elastic region. The progress of elastic to plastic behavior is expressed by the yield strength point. The yield strength is the stress value obtained from the stress–strain curve when intercepted by a straight line parallel to the elastic fraction which is offset by a specified strain (usually from 0.2% to 2%). The tensile strength is the maximum stress value endured by the object in a tension test, and thus the highest point in the strain–stress plot.

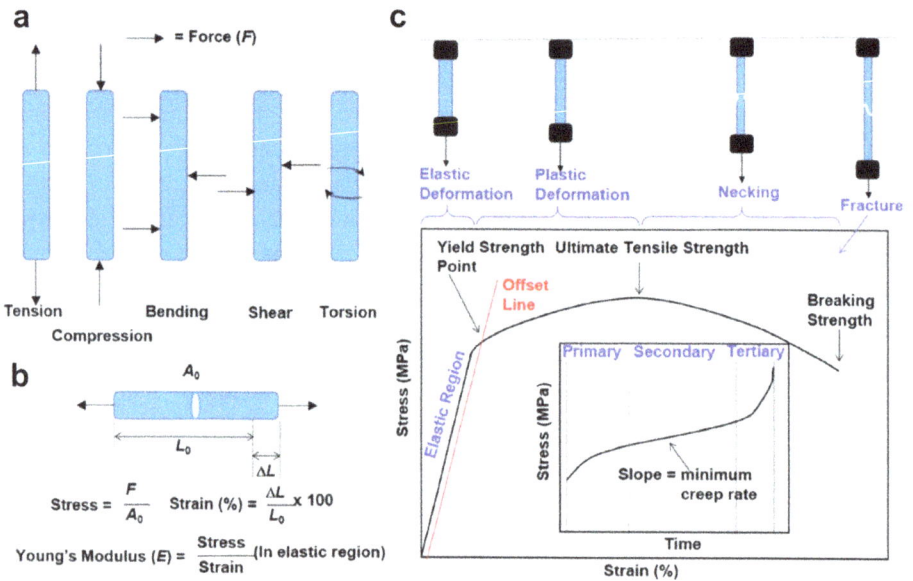

Figure 1.5: (a) Illustration of various loading conditions to a wire. (b) Representing tensile load applied to a wire and the related equations. (c) Representative stress–strain curve for a material and inset shows the time–stress curve to access creep (reproduced with permission from ref. (60)).

1.5 Summary

Here, the advances that have been emerged in design concepts, fabrication methods, and mechanical and electrochemical performances for energy storage devices are described. In order to develop sustainable energy storage systems, their mechanical properties need to be identical with wearable substrates such as skin and fabrics. Hence, it is imperative to develop flexible and stretchable supercapacitors and LIBs. The fabrication of such systems requires flexible and stretchable electrode material, current collector, solid-state electrolyte, and encapsulating material. The architectural design of energy storage devices is typically important, and wire/fiber configurations, origami patterns, and island-bridge designs have shown tremendous potential for wearable applications. A deep analysis of the parameters that describes the bending and stretching state is crucial for rationally designing the structural patterns of the devices. The mechanics of both wavy and island-bridge design for stretchable electronics showed how the buckled profile diminishes the strain to accomplish large stretchability. Both the buckling geometry (wavelength and amplitude) and the peak strains are illustrated analytically. Furthermore, the major developments of utilizing conducting nanomaterials and intrinsically stretchable binders for developing elastomeric conductor are summarized.

2 Flexible and stretchable supercapacitors

2.1 Introduction

In capacitors, the electrical energy is stored directly in an electrostatic way between two parallel plates having positive and negative charges, usually called electric double-layer capacitors (EDLCs). Such non-Faradaic electrical energy storage system does not involve any chemical change, but only an excess and deficiency of charges are developed during charging and the discharging processes. As no phase or chemical transitions are happening, the capacitors usually provide limitless cyclability with negligible capacitance losses (122). Carbon nanomaterials, like carbon nanotubes (CNTs), activated carbons, carbon aerogels, graphene, and carbide-derived carbon are exceptional active materials for EDLCs owing to their large specific surface area, good chemical stability, high mechanical endurance, and great electrical conductivity (123–125). A supercapacitor is also termed as an ultracapacitor which greatly differs from the ordinary capacitor in terms of its higher capacity and energy density, while maintaining a higher power density. Supercapacitors possess high specific capacitance, long cycle life, large power density, and high capacitance retention, are maintenance free and safe, and can function as a bridging power source for buffering the power–energy difference that exists in other energy sources or harvesters like fuel cells, solar cells, or batteries (126). Such characteristics make the supercapacitors a suitable power source for systems that necessitate high power and durability (127).

In order to improve the charge storage capacity of the supercapacitors researchers have been investigated the pseudocapacitive materials having Faradaic charge transfer properties. RuO_2 pseudocapacitors store charge through oxidation-reduction reactions, electrosorption, and intercalation mechanisms (128). Pseudocapacitance is related to the electron transfer at the electrode–electrolyte interface significantly from desolvated and adsorbed ions (129). The ability of electrode materials for pseudocapacitive effect depends on many factors like their chemical affinity to the adsorbed ions and structure and the dimension of the electrode pores (129). Several materials that exhibit redox behavior are widely used in pseudocapacitors such as transition-metal oxides (MnO_2, RuO_2, IrO_2, Fe_3O_4, V_2O_5, NiO, Co_3O_4, etc.) (130–134), transition metal sulfides (135), and conducting polymers like poly(3,4-ethylene dioxythiophene) (PEDOT), polyaniline (PANI), polythiophene, polypyrrole, poly-phenylene-vinylene, poly, and polyacetylene (136–140).

Hybrid capacitors are developed by coupling a supercapacitor electrode with a battery electrode. In order to construct these capacitors, a carbon electrode is usually integrated with a Li-ion electrode, and otherwise called Li-ion capacitor. Such combination enhances the specific capacitance, lowers the anode potential with augmented cell voltage, and thus increases the energy density (141). In these systems, the Faradaic battery type electrode provides higher energy density and the

https://doi.org/10.1515/9781501521287-002

non-Faradaic electrode provides higher power density. The existing and future applications in multifunctional consumer electronics mandate thinner, flexible, lighter, and transparent supercapacitors. Albeit the activated carbon and the transition metal oxides are good electrode materials, challenges are still exist to meet the power requirements (142). The irregular morphologies of the carbonaceous materials and the poor electronic conductivities of the metal oxides limit their application in high-rate energy storage environments. Researchers recently explored novel electrode materials like metal-organic frameworks, covalent organic frameworks, 2D materials, MXenes, metal nitrides, metal sulfides, and mixed conductors (143–145).

The critical challenges in constructing flexible and stretchable supercapacitors are the choice of bendable current collectors with good electronic conductivities and mechanical properties. In general, a supercapacitor device consists of two electrodes (to function as anode and cathodes) interfaced by a separator and a liquid electrolyte with good ionic strength and conductivities. The use of solid gel electrolytes is needed for safe practical applications because the liquid electrolytes face the risk of leakage. Also the solid electrolytes can often eliminate the separator as their ability to avoid electrical contact and short circuits inside the system. The development of flexible electrode materials with excellent specific capacitance and rate capability, and their integration with the flexible current collectors, are crucial to maintain the device performance when subjected to repeated mechanical deformations. The upcoming sections describe the immense research efforts have been made to realize highly flexible supercapacitors for wearable applications.

2.2 Fiber-based flexible supercapacitors

Fibers are usually produced by spinning process; the fiber-forming materials are dissolved or melted to form a thick viscous liquid and then forced through a spinneret having hundreds of nanosized holes. The fibers with different thicknesses can be developed by extruding through the holes having different dimensions. Later electrospinning was evolved as a fascinating technique for fabricating nanostructured fibers. Electrospinning possesses characteristics of both conventional electrospraying and solution dry spinning of fibers. Electrospinning uses electric force to draw charged threads of viscous polymer solutions or polymer melts through the nanosized slits. The process involves the charging of the body of a liquid by applying high voltage to a liquid droplet. When the electrostatic repulsion counteracts the surface tension of the liquid the droplet is stretched and a stream of liquid erupts from the surface at a critical point called the Taylor cone. If the molecular cohesive forces of the liquid are high enough, the stream disintegration does not happen and a charged liquid jet is aroused. The jet flow through the air dries the fiber and charges are migrated to its surface and the form of current flow transforms from ohmic to convective. The electrostatic repulsion is started at small bends in the fiber and the jet is further elongated until it is deposited at the grounded collector. Uniform thinning and elongation of the fiber happen during this bending instability that leads to the formation of homogenous fibers with nanosized diameters (146).

Fiber-like or wire-shaped supercapacitors are suitable for supplying energy for flexible microelectronic devices (147). The advanced weaving technologies facilitate yarn-shaped flexible electrodes having good mechanical strength which can be woven into different types of clothes. This low-cost method offers effortless fabrication of large area devices that could be seamlessly merged with the wearer's daily life without affecting their day-to-day activities. The fiber-based smart cloths that can store energy opens a valuable platform for various wearable applications in healthcare, fitness monitoring, bendable display, and many more. There are three different configurations for fiber like supercapacitors such as parallel, twisted, and coaxial structures. The type 1 structure is prepared by parallelly aligning two functionalized fiber electrodes as anode and cathode, respectively (148). A spacer is needed for parallel supercapacitors to effectively avoid short circuits caused by the direct electrical contact between two electrodes. The spacer must provide good ion transport, so porous thin polymer structures or gel electrolytes itself are generally used. Another one efficient approach is the use of helical spacer wire that could prevent adjacent fiber electrodes from short-circuiting, even during bending or twisting deformations and maintains ion transport. The spacer wire needs to be evenly twisted around one of the conductive fiber electrode with a certain pitch. Then both of the electrodes can be parallelly aligned and packaged inside a flexible tube with electrolytes (Figure 2.1a). In type 2, two yarn-shaped supercapacitor electrodes are twisted around each other, and type 3 stands for a coaxial structure where an inner fiber-like flexible electrode is wrapped by a film-like outer electrode (also called "membrane electrode"). One example of a twisted fiber structure is illustrated in Figure 2.1b, where the carbon fibers are functionalized with activated carbon, coated with the gel electrolyte, and two of such electrodes are twisted each other (149). For the twisted supercapacitor, the direct interface area between the fiber electrodes is limited. During mechanical deformations, such fiber electrodes could physically separate from each other, resulting in gel deterioration, poor ionic conductivity, and high internal resistance. The coaxial structure accommodates bundles of fiber electrodes at the core and thus possesses large surface area, high capacitance, high energy density, and power density. The coaxial configuration involves consecutive deposition of current collectors, active materials, and electrolytes over the fibrous support (Figure 2.1c), and such devices are structurally more robust to withstand the mechanical strain during deformations (147). However, bottlenecks still exist for its scalable production, which are the difficulties related to the precise layer-by-layer deposition of thin films onto the ultrathin long fibers. It is always imperative to eliminate the short-circuit problems in all three types of fiber-based supercapacitors. The fiber supercapacitors can be worn on the neck, wrist, or any other curved region in the human body, and they allow complete freedom in the device design and construction for realizing practical wearable applications.

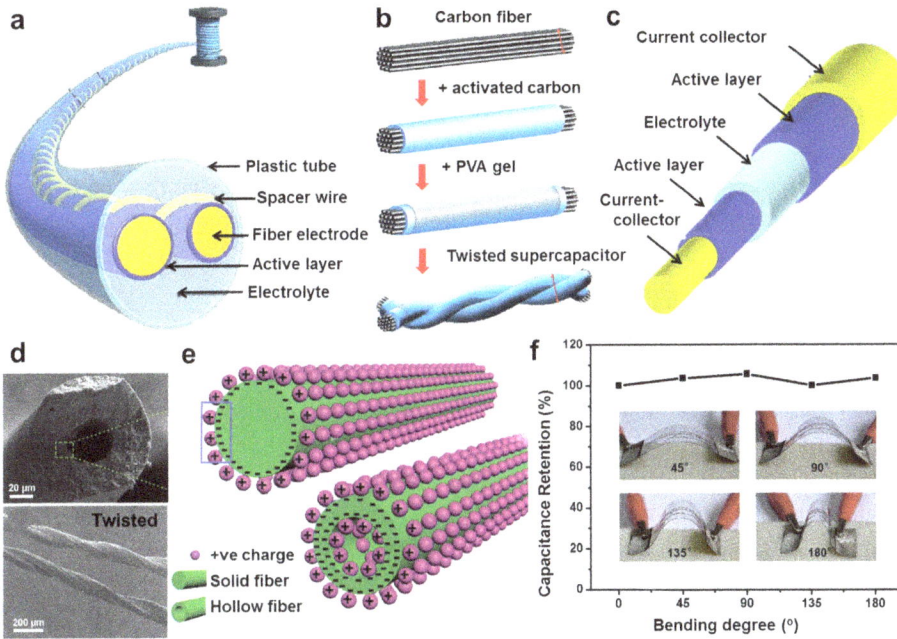

Figure 2.1: Schematic illustration of the (a) architecture of fiber supercapacitor (reproduced with permission from ref. (148)), (b) fabrication process of all-carbon solid-state yarn supercapacitors (reproduced with permission from ref. (149)) and a (c) single coaxial fiber supercapacitor (reproduced with permission from ref. (147)). (d) Cross-sectional SEM images of the hollow composite fiber and two of such twisted fibers. (e) Schematic demonstration of the charge distribution on the solid fiber and hollow fiber capacitor. (f) Optical images of the fiber supercapacitor under different bending states and the corresponding capacitance retention values (reproduced with permission from ref. (150)).

2.2.1 Electrode materials for flexible fiber supercapacitor

The ultrathin fibers are mechanically stable and mostly used as a flexible support or current collector and electrochemically active materials are directly grown on its surface to improve charge storage performances. Different types of fiber electrodes have been developed in the past years, generally exploiting plastic fibers, ultrafine conductive metal fibers, CNT or graphene yarns, polymer fibers, and composite fibers. The plastic and polymer fibers suffer from poor electronic conductivities, and the issue can be addressed by coating active materials like CNTs, activated carbon, graphene nanosheets, transition metal oxides, and conducting polymers. The modification is achieved by using several approaches like dip coating and drying, electrodeposition, electroless coating, and chemical bonding. Albeit the fiber-based electrodes offer scalable production for practical applications, the higher weight or volume fraction of the fibrous support than the electrochemically active materials causes an adverse effect on the specific capacitance. Hence it is important to develop suitable fibrous materials that possess both the functions of mechanical resiliency and charge storage capabilities. Many of the fiber supercapacitors use additional metallic current collectors like Pt wire or Ni mesh for attaining better performances. But the use of conductive fibrous materials can avoid the convoluted assimilation of separate current collectors. The upcoming sections describe the different electrode fabrication approaches attempted for developing fiber-based flexible supercapacitors.

2.2.1.1 Carbon-based flexible fiber electrodes

The carbon-based materials provide an excellent platform for supercapacitor electrodes thanks to their remarkable electric conductivity, thermal and chemical stability, high surface area, low-cost, lightweight, wide operating temperature range, non-toxic, and being easy to operate. The charge storage behavior of the carbonaceous materials is governed by the electrical double-layer capacitance, where the charges are electrostatically stored at the electrode–electrolyte interface through adsorbing–desorbing phenomena. So far, different types of carbon materials have been utilized in fiber supercapacitors, including activated carbon, single-walled carbon nanotubes (SWCNTs) and multi-walled carbon nanotubes (MWCNTs), graphene, ordered mesoporous carbons, and so on. Yang et al. developed ultrafine carbon fiber webs through electrospinning of poly(amic acid) (PAA) solutions (150). The electrospinning process is fast, reliable, and useful for producing fibers with high porosities and large surface-to-area ratios. The PAA fiber was further imidized and finally carbonized in the temperature range from 700 to 1,000 °C under nitrogen atmosphere. The carbonized webs activated at the temperature range of 650–850 °C showed high surface area of 940–2,100 $m^2\ g^{-1}$ and the specific capacitance of 175 F g^{-1} at a current density of 1,000 mA g^{-1}.

The activated carbon is usually in powder form and their large specific surface makes it a common electrode material for electric double-layer-based capacitors. The fiber form of activated carbon is suitable for large-scale fabrication of flexible supercapacitors and the processing needs stable dispersion in solution at a high concentration appropriate for productive coagulation. The use of traditional dispersants like surfactants and polymer causes increased resistivity and thus reduces the performances. Zhu et al. developed activated carbon fibers using a bottom-up method by exploiting graphene oxide (GO) as both dispersant and binder (151). The electrical conductivity of the fiber can be improved by reducing GO into highly conductive rGO and such binder free fiber with conducting additive provide high capacitance of 27.6 F cm^{-3}, good cycling stability even after 10,000 cycles with 90.4% capacitance retention, and good bendability for different bending angles and cycles.

The typical structural, electrical, and mechanical properties of CNTs facilitate to utilize as an electrode materials in planar energy storage systems. The charge transport in CNT materials can be improved by aligning it into fiber-shaped electrodes. Peng et al. developed aligned MWCNT fibers by spin coating from the array of CNTs prepared by chemical vapor deposition (152). The array of CNTs has controlled diameters from 2 to 30 μm and lengths up to 100 m. The flexible wire-shaped microsupercapacitor was fabricated by twisting two aligned MWCNT fibers each other and possessed a mass specific capacitance of 13.31 F g^{-1} and area specific capacitance of 3.01 mF cm^{-2}. Kim et al. reported a flexible CNT yarn electrode and further coated with MnO$_2$ (153). The yarn electrode is prepared by twisting MWNT sheets to allocate high internal 3D porosity and MnO$_2$ was trapped inside the pores during deposition. The uniaxial aligned MWCNT bundles of yarn was employed as current collector, and the final all-solid-state yarn supercapacitor provides high degrees of specific capacitance (25.4 F cm^{-3} at 10 mV s^{-1}) and energy densities (3.52 mWh cm^{-3}).

Graphene possesses unique characteristics such as high surface area, good chemical stability and electrical conductivity, and hence it is an excellent electrode material for developing high-performance supercapacitors. Graphene fibers exhibit enormous strength and electrical conductivities, mechanical flexibility, and light weight. The large-scale constructions of graphene fiber are possible through hydrothermal assembly and wet spinning of GO (154). Qu et al. developed a fiber supercapacitor via region-specific laser reduction of GO fiber forming a controlled configuration of rGO–GO–rGO, where rGO sheets function as electrodes and GO serve as the separator and are sensibly integrated in single graphene fiber (155). Gao et al. developed coaxial type flexible fiber electrodes by growing NiCo$_2$O$_4$ nanosheets on Ni wire and they showed good capacitance retention during repeated cycles of mechanical deformation.

2.2.1.2 Metal oxide–based flexible fiber electrodes

Transitional metal oxides are promising candidates for developing flexible supercapacitors owing to their remarkable properties such as high theoretical specific capacitance, large surface area, morphologies, environmental friendliness, abundant raw materials, and easy synthesis approaches. The capacitance of metal oxides can be prominently improved by controlling and varying their defects at the surface. Even though the metal oxides provide high energy density, their poor electrical conductivity and unmanageable volume expansion and sluggish diffusion of ions in the bulk phase have limited their realistic applications. Therefore, it is important to explore functional metal oxide materials with significant charge storage properties in terms of their electronic conductivity, composition, and oxygen vacancies. The coexistence of two different cations in a metal oxide structure could generate more number of electrons than in a single metal configuration and thus it augments the electrical conductivity (6).

Wet-spinning method is a scalable method for preparing highly conductive two-dimensional (2D)-MnO_2 nanosheets loaded SWCNTs fibers (156). The 2D structure offers highly active interfaces, fast electrode kinetics, and a shortened ion-diffusion length during charge-discharge process. MnO_2 possesses large theoretical capacity of ~1,370 F g^{-1} and provides fast charge/discharge kinetics over a broad-range of potential window. The SWCNT serves as a conducting structure, flexible mechanical support and spacer for the dispersed MnO_2 nanosheets and the as prepared flexible hybrid fiber supercapacitor showed very high volumetric capacitance of 74.8 F cm^{-3} and an energy density of 10.4 mWh cm^{-3}.

All-graphene core sheath fiber provides high electronic conductivity, where a core of graphene fiber is covered with a 3D porous graphene network. The large surface area of 3D graphene provides exceptional energy storage properties, and the core sheath fiber can be easily woven into textile substrates and possess really compressible and stretchable properties. Qu et al. developed MnO_2-loaded graphene sheets on graphene fiber as a core-sheath structure, and the two of such unique fibers are intertwined and solidified in the polyvinyl alcohol (PVA)–H_2SO_4 gel electrolyte to get all-solid-state fiber supercapacitors (157). Such amalgamation of MnO_2 nanostructures onto the large surface of graphene sheets revealed enhanced specific capacitance. Bismuth oxide nanotubes showed good electronic conductivity, specific capacitance, and environmental friendliness which makes them good platform for charge storage applications. Gao et al. incorporated the bismuth oxide nanotubes to improve the capacitance of graphene fiber-based flexible supercapacitor prepared by wet-spinning method (158). This method opens an exciting avenue to extend to other metal oxides for productive real-time applications. Other metal oxides such as RuO_2 (159), V_2O_5 (160), and Fe_2O_3 (161) and Co_3O_4 (162) are also widely studied for flexible supercapacitor applications.

2.2.1.3 Conducting polymer-based flexible fiber electrodes

Conducting polymers are the organic polymers that conduct electricity via conjugated bonds along the polymer system. The conducting polymers are promising platform for charge storage application owing to their reversible Faradaic nature, high electrical conductivity, high charge storage capacity, and low-cost. Polymers like polyaniline, polypyrrole, polythiophene, and their derivatives have been extensively explored due to their high pseudocapacitance, flexibility, and ease of production. The nanostructured conducting polymers like nanosheets, nanowalls, and nanorods provide high surface area, porosity, and high surface-to-volume contribution; hence, these nanostructures could impart high pseudocapacitance compared to their bulk materials (163).

Significant research developments have been achieved through the use of carbon-based fiber electrodes and their composites with conducting polymers. A thread-like supercapacitor is developed from CNT yarns and polyaniline nanowires (164). The supercapacitor possesses very good charge storage capacity and when woven or knitted into wearable textile substrate, the device can be bended and stretched repeatedly without affecting its charge storage ability. The capacitance of the fiber electrodes can be further augmented by increasing the interfacial area to efficiently interact with the electrolyte. If the fiber electrode is hollow in nature, the inner region of the electrode is also available to contribute for capacitance with the help of higher surface area. The larger inner hollow area of the fiber offers substantial specific surface area, thus facilitating augmented electrode/electrolyte contact region and higher charge transfer kinetics. For instance, Peng et al. developed such a hollow fiber electrode by utilizing hollow composite fibers of rGO and poly(3,4-ethylenedioxythiophene):poly(styrenesulfonate) (PEDOT:PSS) (165). The scanning electron microscopy (SEM) image of a cross-sectional composite fiber electrode shows an apparent interior hole, and the fibers could be twisted each other owing to their good mechanical strength (Figure 2.1d). As shown in the schematic, the solid fiber electrode contributes the capacitance only from its outer surface, but in hollow fiber, the positive charges are distributed at both the outer and inner surfaces and the corresponding charge separation occurs within the fiber, resulting in enhanced charge storage properties (Figure 2.1e). Such symmetric hollow fiber supercapacitors having PVA/H_3PO_4 gel electrolyte possess high charge-storage performance (304.5 mF cm^{-2}) and good capacitance retention during bending deformations (Figure 2.1f).

Polypyrrole is a conducting polymer that holds several advantages like high Faradaic capacitance, environmental stability, and good electronic and ionic conductivity. Challenges are existing to develop polypyrrole fibers from its powder form and its over oxidized form suffers from poor conductivity In order to prepare a stable polypyrrole (Ppy) fibers with improved mechanical flexibility and charge transfer kinetics a conductive support is usually required. A hierarchically interconnected PEDOT:PSS/PPy fiber-based hybrid supercapacitor electrode was developed by in situ chemical polymerization of polypyrrole on PEDOT:PSS hydrogel fiber (166). The hydrothermally assembled PEDOT: PSS fiber is highly porous and conductive, and the hybrid fiber possesses strong π-π interaction that facilitates rapid electron transfer or ions diffusion throughout the device.

2.2.1.4 MXene-based flexible fiber electrodes

MXene is a novel family of 2D transition metal carbides and nitrides that can function as electrodes, interconnects, and current collectors due to their excellent electrical conductivity (6.8×106 S m^{-1} for pure $Ti_3C_2T_x$ film), high specific area, and robust mechanical tunability (167). MXenes possess the general formula $M_{n+1}X_nT_x$, where M represents a transition metal (n = 1–4), X indicates carbon and/or nitrogen, and T_x is the surface terminal groups (–O, –OH, –Cl, –F). The thin film of MXenes at optimal synthesis conditions showed very high conductivity of 15,000 S cm^{-1} (168) and volumetric capacitance of 1,500 F cm^{-3} (169). The hydrophilic nature of MXene provides good solution processability even in the absence of surfactants, and its ability to 3D print, spray-coat, and spin-coat has been demonstrated (170–172). These properties make MXene a promising electroactive material for constructing flexible fibers for smart electronic textile application.

MXene-based fibers are in need for energy storage in portable and wearable intelligent devices. Unlike polymers, MXene sheets are unable to process molten form because it can be easily oxidized before melting. Wet-spinning technique is suitable for preparing ordered fiber microstructures via extruding spinning solution through the nozzle, followed by soaking the treatment. Gao et al. utilized this wet-spinning assembly strategy for preparing very long MXene-based fibers by means of a synergistic effect between GO as a liquid crystalline additive and MXene sheets (173). The hybrid fiber possessed uniformly aligned microstructures with appreciable conductivity of 2.9×10^4 S m^{-1}. The assembled yarn-shaped symmetric supercapacitor showed an excellent charge storage performance with volumetric capacitance of 586.4 F cm^{-3}.

The direct coating of Mxene on fiber structural units is widely explored for wearable applications (174). There are some challenges associated with the coated fibers, especially the flaking off the active materials from the fiber during bending, twisting, weaving, or knitting process that could adversely affect the electrochemical performances. Also, the excess loading of active materials causes restacking of MXene sheets, which restricts the mobility of ions and thus the capacitive behavior. Hence it is important to develop MXene fiber composites by embedding the MXene flakes into fibrous materials. As discussed previously, the electrospinning is a versatile technique for producing nonwoven mates of nanomaterials incorporated fibers with tunable properties. Such freestanding composite fibers are highly porous in nature and provide enhanced access of electrolytes and increased adsorption of ions. The limiting factor for electrospinning of 2D materials is the possibility of agglomeration and restacking of sheets. Hence, it is important to use a stable homogenous suspension of MXene for electrospinning of fiber composites. Gogotsi et al. exploited the electrospinning method to construct $Ti_3C_2T_x$ embedded carbon fibers while maintaining the MXene structures during the carbonization process (175). These freestanding fibers demonstrated augmented charge storage performances in comparison to pure carbon fibers, and are stable during bending folding and twisting deformations.

The good electrical conductivities of the MXene materials facilitate to scale up the fiber electrodes from centimeters to meters without affecting their performances. Table 2.1 shows the recent advances associated with the MXene-based fiber, yarn, and fabric supercapacitors (173, 175–187). The excellent mechanical and electrochemical properties of these materials are suitable for producing of woven and knitted energy storage devices for powering textile-based wearable electronics.

Table 2.1: The electrical, mechanical, and energy storage properties of MXene-based fibers, yarns, and fabrics.

Electrode type	Electrode	Fabrication method	Conductivity S cm^{-1}	Tensile strength Mpa	C_v F cm^{-3}	C_a mF cm^{-2}	References
Fiber-based electrode	MXene/ PEDOT:PSS fiber	Wet spinning	1,489	58.7	615	676	(176)
	MXene/ CNT@PCL fiber mat	Electrospinning	N/A	3	N/A	50	(177)
	MXene/ carbon fiber mat	Electrospinning	N/A	N/A	N/A	244	(175)
	MXene/ rGO fiber	Wet spinning	290	12.9	891	565	(173)
	MXene/ rGO fiber	Wet spinning	72.3	132.5	341	233	(178)
Yarn-based electrode	MXene/ PEDOT:PSS coated Carbon tow	Coating	198	3,000	N/A	659	(179)
	Biscrolled MXene/ CNT yarn	Biscrolling	2.7	38.4	92	N/A	(180)
	Biscrolled MXene/CNT yarn	Biscrolling	N/A	26.6	1,083	3,188	(181)
	MXene/ silver-coated nylon yarn	Coating	N/A	N/A	N/A	328	(182)
	MXene/PET/ silver-coated nylon yarn	Electrospinning	N/A	N/A	72	N/A	(183)
	MXene-coated cotton yarn	Coating	199	460	0.26	3,965	(184)

Table 2.1 (continued)

Electrode type	Electrode	Fabrication method	Conductivity S cm^{-1}	Tensile strength Mpa	C_v F cm^{-3}	C_a mF cm^{-2}	References
Fabric based electrode	MXene-coated carbonized silk fabric	Coating	N/A	N/A	362	N/A	(185)
	MXene-coated cotton knit fabric	Coating, knitted	N/A	N/A	N/A	707	(186)
	MXene-coated carbon fiber fabric	Coating	N/A	N/A	N/A	416	(187)

C_v – volumetric capacitance; C_a – areal capacitance

Despite the fact that immense research developments and advances are associated with flexible fiber supercapacitors, there are still unresolved problems to be addressed. The energy storage performance of fiber supercapacitors are far behind than the planar devices, which hinders their practical applications. The energy densities and capacitance of fiber supercapacitors need to augment while maintaining their inherent high specific power densities. It is crucial to increase the charge-storage capacity and decrease the volume of the fiber devices, which provides sufficient energy density to power wearable devices. In addition, several versatile approaches must be developed for producing cost-effective fiber substrates on a large scale. The current approaches are limited to materials like CNTs, graphene, and their composites with conducting polymers. Metallic fibers are another efficient class of active substrate owing to their electrical conductivity and mechanical strength. But the diameters of the reported metallic fibers are generally very big, which increases the volume and reduces the flexibility of the supercapacitors. Hence, fabrication of thin metallic fibers with adequate flexibility, tensile strength, and charge storage properties is crucial for developing self-powered wearable devices.

2.3 Stretchable fiber supercapacitors

Majority of the fiber-shaped supercapacitors possess outstanding flexibility and remains unaffected when exposed to external stress. However, when the applied stress is increasing beyond a limit, crack formation occurs and badly affects its conductivity and capacitance. If the fiber possesses stretchable properties, it can withstand enormous strain developed during harsh mechanical deformations. Therefore the retention of charge storage abilities during stretching of the fiber supercapacitor is crucial for their productive incorporation into textiles.

Peng et al. developed a fiber-shaped stretchable supercapacitor based on aligned CNT sheets which are sequentially wrapped around an elastic fiber (188). The fabrication process of such coaxially stretchable device includes the successive coating of gel electrolyte and wrapping of the CNT sheets, respectively, around the rubber fiber (Figure 2.2a). The aligned CNT sheets provide high flexibility, electrical conductivity, tensile strength, and thermal and mechanical stability. The fibers can be wrapped on substrates having different shapes (Figure 2.2c), and the can be stretched into 100% of its initial length without any mechanical deformations (Figure 2.2c). During stretching, the gel electrolyte (PVA-H_3PO_4) stabilizes the CNTs and the wrapped CNT sheets retain the aligned structure similar to springs. As both the elastic fiber and gel electrolyte are stretchable, the supercapacitor sustains its high performance (18 F g^{-1}) when it is exposed to various levels (maximum 75%) of stretching cycles. As MXene provide excellent conductivity and charge storage property, developing highly flexible MXene fiber supercapacitor is critical in some situations like knitting and stretching. Xu et al. developed a core-sheath asymmetric fiber supercapacitor based on $NiCo_2S_4$-loaded CNT film as an outer positive electrode and Ti_3C_2Tx-decorated carbon cloth fiber as a core negative electrode (189). The all-in-one supercapacitor twisted into a helicoid configuration provides reversible stretchability for 20% strain and maintains its performances under repeated bending and stretching cycles.

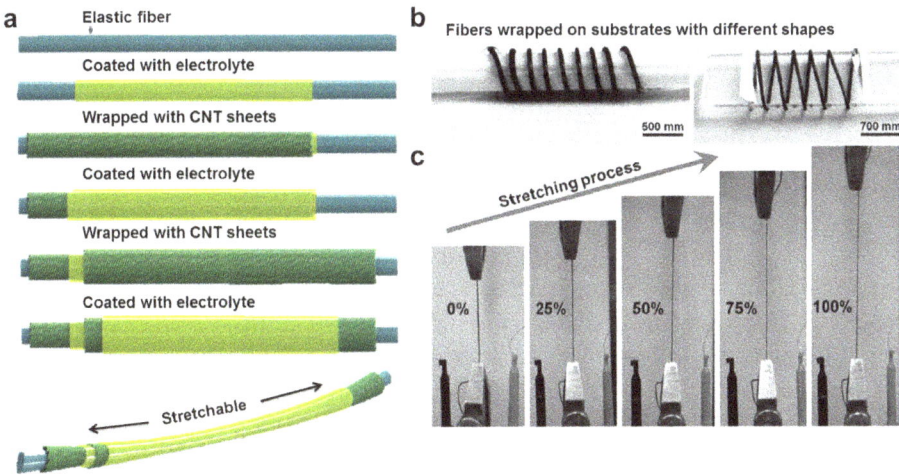

Figure 2.2: (a) Schematically showing the fabrication of a stretchable, fiber-shaped coaxial supercapacitor. (b) Photographs representing (b) two fiber-shaped supercapacitors wound on substrates having different shapes and (c) the fiber supercapacitor under different strains of 0%, 25%, 50%, 75%, and 100% (reproduced with permission from ref. (188).

Common yarns prepared from majority of polymers can sustain only limited strain which is below the degree of deformation of yarn-based stretchable supercapacitors. Hence in order to achieve stretchability these yarns need to be knitted/weaved into fabrics. The stretchability of as produced textile is associated with the interweaving topology that leaves gaps for deformation. The size of the gaps controls the stretchable nature and the fabrics can be stretched maximum of 100% strain. Hu et al. developed urethane-based core spun yarns constructed by interweaving cotton yarns around urethane filaments (190). The fiber offers a promising support material for decorating CNTs and Ppy by simple dip coating or electrodeposition pathways. This methodology provides exceptional mechanical properties to withstand strain of 80% with an areal capacitance of 69 mF cm^{-2}.

Fiber supercapacitors possess very small volume and enormous flexibility which makes them to integrate with vast variety of wearable substrates. Compared with other 2D or 3D supercapacitors, fiber devices can be utilized to power miniaturized portable electronic devices, textile-based sensors for healthcare monitoring, and implantable bio-integrated medical devices. But majority of fiber supercapacitors are suffering from low specific energy, and unable to be effectively used as a power source. As the stored energy in the capacitor is proportional to the specific capacitance and the operating potential window, the energy density can be augmented by enhancing these two factors (191). As discussed previously, the incorporation of electrochemically active components like nanosized metal oxides, 2D materials, and conducting polymers greatly improves the charge storage abilities. The voltage range extension for fiber supercapacitors could deliver sufficient power for wearable electronics applications. The most productive methodology is the integration of asymmetric supercapacitor having different positive and negative electrode species with well-separated potential ranges. Several studies are reported for such asymmetric fiber supercapacitors having tremendous mechanical flexibility and wide operating voltage (192–196). Introducing stretchable properties to the asymmetric fiber electrodes could further improve the mechanical properties of the device for real wearable applications.

A coaxial asymmetric supercapacitor was developed with stretchable properties by Chou et al. that used MnO$_2$ nanostructures modified CNT fiber as a positive electrode and CNT-polypyrrole composite film as a negative electrode (192). The presence of CNTs and polypyrrole greatly boosted up the capacitive nature and widened the operating potential from 0.8 to 1.5 V. The coaxial structure provides another opportunity of utilizing maximum of effective interfacial area of both the electrodes, thus enabling high charge storage for supercapacitors. A helical configuration with uniform loops is structured by over-twisting the core-sheath supercapacitors, which helps to realize stretchable freestanding fibers. The supercapacitor maintained its capacitive nature during 20% stretch-release cycles and provides high energy and power densities.

Studies on stretchable fiber supercapacitors are still at the beginning stage and immense research needs to focus on developing stretchable fiber substrates. More competent manufacturing techniques are required to produce integrated hybrid systems with compact size, reduced weight, and improved performances. At present, fiber-based supercapacitors are woven either by hand or by tiny equipment on a small area. Immense research is focused on to exploit the industry-level machines for weaving fiber supercapacitor on large areas of wearable fabric substrates. The stretchable fibers provide mechanical strength and stability similar to the usual fibers used in commercial textile industry. In order to establish woven fiber-based energy storage devices, reliable materials must be judiciously chosen for current collector, electrode, electrolyte, and packing source.

2.4 Intrinsically stretchable supercapacitors

Intrinsically stretchable supercapacitors consist of completely stretchable components without depending on any configuration- or structure-based strain management. The development of such supercapacitors are difficult to be achieved because the stretchability to be incorporated to the electrode material of the device. These devices can be prepared by either utilizing intrinsically stretchable electroactive materials or by making an elastomeric composite of active materials and a stretchable binder. The loading of active materials needs to be optimized to have sufficient charge storage behavior and appreciable conductivity without compromising its stretchable properties and mechanical resiliency. The intrinsically stretchable systems can other distinct benefits like low cost of production and isotropic strain ability. The challenging factors associated with this type of supercapacitors are the requirement for stretchable counterparts including gel or separator. The unsymmetrical stretchability of different components causes interfacial delamination or misalignment during repeated stretch-release cycles (197).

The initial development in the field of stretchable supercapacitor was focused on coating a layer of active materials onto the intrinsically stretchable conductive substrate. Tong et al. reported a water-assisted synthesis of CNT-PDMS composite as an intrinsically stretchable current collector film (198). The film containing 10% CNTs possessed appreciable conductivity of 4.19 S cm^{-1}, and able to stretch to a limit of 50% strain. The CNT-PDMS film coated with polyaniline nanofibers possessed a specific capacitance of 1,023 F g^{-1} at a scan rate of 5 mV s^{-1}, and 95% capacitance retention during 500 cycles of dynamic stretch-release cycles. However, challenges exist in the mechanical properties of the non-stretchable layer of active materials on top of a stretchable conductive electrode. The judicious synthesis of shape conformable polymer composites with active materials offers unique flexibility and stretchability. Acrylate polymers are well known for their resistance to break, elasticity, transparency, and widely used as an adhesive (199). Wang et al. employed intrinsically stretchable acrylate rubber for constructing CNT-based conductive composite film by chemical crosslinking method. The film with 35%

CNT loading withstands 55% of strain and hold good conductivity (9.6 S cm^{-1}). The conductive path facilitates electrodeposition of conducting polymers for developing stretchable supercapacitors with high energy densities. Lu et al. developed a ternary composite of ethylvinylacetate with polyaniline and CNTs (200). The composite possesses a 3D co-continuous phase structure and an elongation at break of 55%, and viable to be molded to any desired shape. Also, the elastic nature of the composite minimizes the structural breakdown of polyaniline during the continuous charge-discharge process.

The block copolymers are the elastomers formed via physical cross-linking and can be easily processed in organic solvents. The viscoelastic and thermoplastic nature of block copolymers enables to function as a soft and stretchable binder for configuring wearable electronic devices (201). As the pore size of electrode components determines the ion-accessible surface area, charge storage ability, rate capability, energy, and power density, the degree of pore formation in block copolymers greatly influences the capacitance of supercapacitors (202). Therefore, the blends of block copolymer and pore-forming additive along with the active materials like activated carbon or mesoporous carbon form stretchable supercapacitors with enhanced performances (43, 203).

2.5 Design-induced stretchable supercapacitors

2.5.1 Pre-strain approach

A very simplest methodology for realizing stretchable system is the pre-strain approach. Basically, a certain level of strain is usually applied to the stretchable substrate and the active materials are further deposited on it. The strain can be applied either by thermal (204) or by mechanical actuation (205). The releasing of pre-strain enables the formation of buckled wavy structures with certain periodicity and which provides certain stretchability to the active electrode film. The carbon-based active materials such as CNTs and graphene nanostructures are well exploited for preparing buckled supercapacitors. The high surface area and electrical conductivities of carbonaceous films facilitate use as electroactive charge storage materials and the current collectors.

Jiang et al. reported a buckled supercapacitor based on a 2D network of sinusoidal-shaped SWCNT films on an elastomeric PDMS substrate (206). The processes involve the pre-stretching of PDMS followed by the surface pre-treatment to get a hydrophilic surface. The UV light illumination creates activated atomic oxygen species on the PDMS surface that facilitates to form strong chemical bonds with inorganic films featured with hydroxyl moieties. Removing the strain from the substrate provides the spontaneous patterning of periodically buckled SWCNT films due to the difference in mechanical properties of the thick substrate and the conductive CNT film. A stretchable supercapacitor is constructed by assembling two of the sinusoidal

CNT films interfaced with an electrolyte and a separator. The supercapacitor possesses remarkable charge storage performance and remains intact when exposed to 30% applied strain.

In order to meet the requirement for completely stretchable integrated devices that may require highly stretchable supercapacitors (preferably above 100%), the electrode configuration and the wavelength of the buckled structure need to be optimized. Even though the CNT films provide large fault tolerance via multiple current pathways, the missing of some interconnects within the film affects its electrical conductivity. The strain-withstanding ability of the buckled configuration is related to its wavelength, and increasing the pre-strain offers enhanced stretchability. Growing SWCNT films directly on the pre-strained elastomer provides buckled system with large strain endurance and tremendous mechanical properties (Figure 2.3a). These films are different from the films prepared by other deposition methods that may hold irregularly oriented structures. The continuous reticulate structures prepared via direct grown approach possess exceptional conductivity of 2,000 S cm^{-1} and higher strain resilience. Xie et al. developed such highly stretchable supercapacitors having 120% strain limit and the performances remains constant during the stretch-release cycles (207).

A major limitation existing in majority of the developed supercapacitors is their inability to stretch in all directions. All the epidermal electronics require wearable power sources that can be stretched to multiple directions so as to mitigate the strain developed during irregular skin movements. Hence an effective approach is required to construct an omnidirectionally stretchable supercapacitor that can hold high degrees of strain and provide high capacitance and energy density during multidirectional stretching deformations. Isotropically buckled CNT architectures offer high levels of omnidirectional stretchability of 200% strain, conductivity of 2.75 Ω sq^{-1}, and mechanical durability (208). Polyaniline was electrochemically deposited on the buckled CNT layer and that further improves the performance by pseudo-capacitive effect. The buckled CNT films with different stretchabilities can be prepared by varying the pre-stain of the substrate and thus controlling the wavelength of the buckled structures. Notably, the buckle formation is isotropic and uniform when the applied strain is released simultaneously in all directions, and non-uniform, anisotropic buckles are formed when the pre-stain is initially removed in one direction and then from rest of the directions. Also, the reversible nature of the buckling-unfolding process enables retention of the high degrees of capacitive behavior even under large cycles of deformations.

2.5.2 Island-bridge method

Microsupercapacitors are one class of planar supercapacitors that allow facile integration in to flexible electronics due to their reduction in size and thickness. These supercapacitors require no separators to function where both the anode and cathode is structured parallelly in a same substrate. The 2D planar configuration reduces the ionic diffusion paths, and thus exhibits high power density and long durability. Island-bridge concept is widely utilized to realize stretchable microsupercapacitors by using serpentine-shaped interconnects between active supercapacitor electrodes. When the supercapacitor is stretched, the unwinding of serpentine pattern mitigates majority of the device strain and renders the active area unaffected.

Ha et al. adopted a strategic design for realizing a stretchable microsupercapacitors array using completely solid state materials (209). The electrodes are patterned by spray coating array of 2D electrodes and coated with triblock copolymer electrolyte, dispersed in an ionic-liquid. The active electrodes are interconnected by serpentine-shaped bridges in a mechanical neutral plane that offers stretchability to a limit of 30% (Figure 2.3b). The interconnects are patterned via various photolithographic processes that offers precise fabrication of high-resolution patterns of nano-sized geometries. The stretchable capacitor exhibited a capacitance of 100 μF at a scan rate of 50 mV s^{-1}, and the performances are stable during 30% stretching deformations.

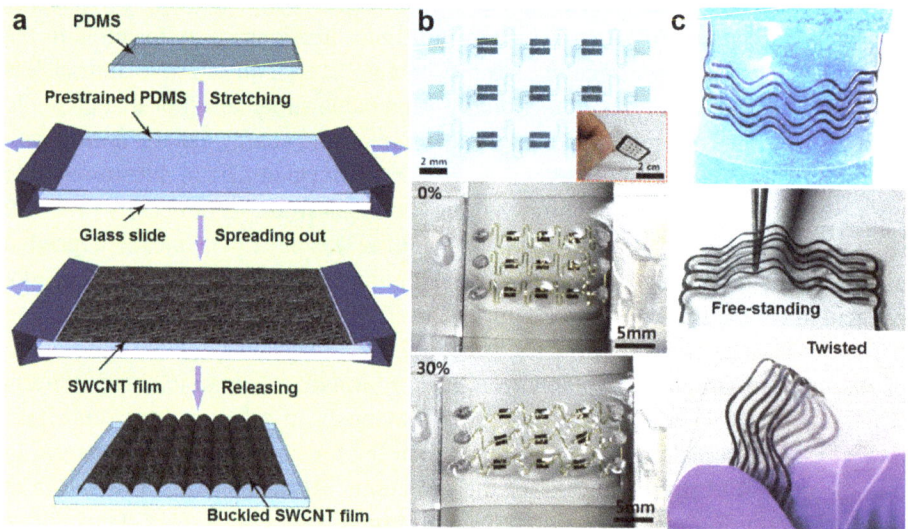

Figure 2.3: (a) Schematic representation of preparing buckled SWCNT film based on pre-strained PDMS (reproduced with permission from ref. (207)) (b) Optical images of a stretchable 2D planar micro-supercapacitor array on a PDMS substrate (inset shows the tape-transferred micro-supercapacitor array) and 3 × 3 micro-supercapacitor array under strain from 0% to 30% (reproduced with permission from ref. (209)). (c) Images showing a wearable serpentine-shaped interdigitated supercapacitor incorporated on a stretchable sweat band, its freestanding nature and flexibility for 180° bending deformation (reproduced with permission from ref. (191).

Although the photolithography generates uniform nanopatterns with high yield and flexibility, the complicated processes are time-consuming, expensive, and depends on clean-room facilities (210). These drawbacks limit its application for large-scale production of multiple identical patterns. This mandates the need for a simpler patterning technique to realize stretchable supercapacitor arrays for wearable applications. The recent development of 3D printing offers an attractive, eco-friendly method for rapidly producing microstructures without depending on a sophisticated manufacturing facility (211). Shen et al. exploited the advantages of 3D printing techniques to develop wavy electrode structures (212). A mold is prepared first followed by casting stretchable PDMS and cured. Further arrays of microsupercapacitor electrodes were prepared by injecting slurry of active materials. MWCNTs functionalized by polyaniline nanorods offer large specific surface area, good conductivity, and high areal capacitance of 44.13 mF cm^{-2}. The charge storage abilities are almost unaltered when the device is stretched from 5% to 40%, and it could power a red light-emitting diode during various mechanical deformations like twisting, winding, and crimping.

The combination of design induced stretchability and intrinsic stretchability is another suitable approach for realizing stretchable supercapacitors. Mohan et al. judiciously developed serpentine-shaped electrodes based on elastomeric polyurethane binder-based stretch enduring inks (191). The stretchable ink consists of CNT-polyaniline as electrochemically active materials, and the electrodes are realized by versatile screen-printing technique. But the direct printing of these interdigitated serpentine electrodes onto the elastomeric substrate is not suitable as it restricts the free unwinding of the serpentine bridges. The freestanding nature of serpentine configuration is crucial to attain the complete advantage of the design-induced stretchability. The freestanding interconnects are realized by printing the desired configuration on a paper substrate having a water soluble (sacrificial) coating, and then transferred onto an elastomeric silicone rubber substrate while bonding only at the island regions. The stretchable supercapacitor showed an excellent areal capacitance of 167 mF cm^{-2} and remains unchanged during repeated cycles of bending and twisting deformations, and the device can withstand a tensile strain of 30% (Figure 2.3c).

Stretchable supercapacitors possess great advantages for skin-conformable wearable applications, especially for powering biointegrated and implantable devices. Table 2.2 discloses various flexible and stretchable supercapacitors developed through different approaches for textile-based, skin-exposed, skin-worn, and implantable applications (190, 213–238). As the human skin shows only limited stretchability (~30%), the exceptional stretchability of these supercapacitors offers robust performances and can withstand extended strain during intense physical activities. In addition, these power sources maintain their performances in many textile substrates, which facilitate realizing cloth-, arm band–, head band–, socks-, or even undergarments-based energy storage devices, for various healthcare, environmental monitoring or defense-related applications.

Table 2.2: Various flexible and stretchable supercapacitors for textile-based, skin-exposed, skin-conformable and implantable applications.

Application	Position	Materials	Configurations	Stretchability	Capacitance	Energy density	References
Textile	Wrist band	SWCNTs	Interdigitated arrays	275%	0.15 F cm^{-3} @ 0.05 V s^{-1}	0.18 mWh cm^{-3}	(213)
	Clothes	Ni-NiCoP/Ni-NiCoP-SWCNT	Textile	100%	877.6 mF cm^{-2} @ 5 mV s^{-1}	40 mW cm^{-2}	(214)
	Clothes	PPy-CB	Sandwich	100%	2.4 mF @ 10 mHz	N/A	(215)
	Clothes	Carbon fiber	Textile	50%	0.51 F cm^{-2} @ 10 mV s^{-1}	N/A	(216)
	Clothes	PPy-CNTs-Urethane	Textile	130%	69 mF cm^{-2} @ 5 mV s^{-1}	6.13 µWh cm^{-2}	(190)
Skin exposed	Glove	AuNWs[a]/Au Film/PANI	Wavy fiber	360%	16.8 mF cm^{-2} @ 0.14 mA cm^{-2}	N/A	(217)
	Glove	CNT	Bulked fiber	200%	11.88 mF cm^{-2} @ 10 mA s^{-1}	5.5 µWh cm^{-3}	(218)
	Glove	SWCNT/PEDOT	Bulked fiber	100%	53 F g^{-1} @ 1 A g^{-1}	6 Wh kg^{-1}	(219)
	Glove	CFT-PANI/FCFT	Wavy fiber	100%	4.8 F cm^{-3} @ 1.6 V	2 mWh cm^{-3}	(220)
	Wrist band	SWCNTs	Interdigitated arrays	275%	0.15 F cm^{-3} @ 0.05 V s^{-1}	0.18 mWh cm^{-3}	(221)
	Watch strap	PANI-CNT	Kirigami	140%	340.2 mF cm^{-2}@2mAcm^{-2}	0.701 mWh cm^{-3}	(222)
	Watch-band	CNF[c]-PEDOT:PSS-CF	Fiber arrays	100%	N/A	85 mW m^{-2}	(223)

Table 2.2 (continued)

Application	Position	Materials	Configurations	Stretchability	Capacitance	Energy density	References
Skin-worn	Wrist/neck	WNT	Interdigitated arrays	50%	4.7 F cm^{-3} @ 0.23 A cm^{-3}	1.5 mWh cm^{-3}	(224)
	Wrist	WO_3-PEDOT:PSS	Sandwich	40%	471 F g^{-1}@1 V s^{-1}	52.6 Wh kg^{-1}	(225)
	Wrist	PPy-CNTs	Serpentine arrays	30%	5.17 mF cm^{-2} @ 100 μA cm^{-2}	0.44 μWh cm^{-2}	(226)
	Finger joint	LIG-N-PEDOT	Interdigitated	400%	790 μF cm^{-2} @ 50 μA cm^{-2}	N/A	(227)
	Forearm	MXene	Kirigami	20%	23 mF cm^{-2} @ 0.1 mA cm^{-2}	2.8 mWh cm^{-3}	(228)
	Hand	Graphite	Kirigami	100%	1 mF cm^{-2} @ 10 mV s^{-1}	N/A	(229)
	Hand	MWCNTs	Interdigitated arrays	100%	0.51 mF cm^{-2} @ 0.006 mA cm^{-2}	0.34 μWh cm^{-2}	(230)
	Arm/knee	$MWCNT/Mn_3O_4$	Interdigitated arrays	50%	8.9 F cm^{-3} @1.5 A cm^{-3}	1.8 mWh cm^{-3}	(231)
	Skin	$CNTs\text{-}RuO_2$	Wavy planar	30%	7 mF cm^{-2} @ 0.5 mA cm^{-2}	N/A	(232)
	Skin	v-AuNWs/PANI	Ultrathin planar	N/A	11.76 mF cm^{-2} @ 10 mV s^{-1}	0.71 μWh cm^{-2}	(233)
	Skin	EG^d/PEDOT:PSS	Interdigitated	N/A	5.4 mF cm^{-2} @ 1 mV s^{-1}	N/A	(234)
	Skin	SWCNT/PEDOT	Ultrathin planar	20%	56 F g^{-1} @ 1 A g^{-1}	6 Wh kg^{-1}	(235)

Table 2.2 (continued)

Application	Position	Materials	Configurations	Stretchability	Capacitance	Energy density	References
Implantable	Mouse tissue	PANI-carbon	Fiber	N/A	4 mF mm^{-2}	450 μJ mm^{-2}	(236)
	Physiological fluids	CNT	Fiber	N/A	10.4 F cm^{-3} @ 0.5 A cm^{-3}	N/A	(237)
	Abdominal muscle of mouse	PEDOT:PSS/ferritin/ MWCNT	Fiber	N/A	32.9 F cm^{-2} @ 10 mV s^{-1}	0.82 μWh cm^{-2}	(238)

[a]Gold nanowire; [b]Carbon fiber thread; [c]Carbon nanofiber; [d]Ethylene glycol

2.6 Summary

The chapter outlined the progress in innovative electrode materials including intrinsically soft and stretchable materials for the development of supercapacitors. In order to realize conformal integration to soft biological tissues having curved and irregular interfaces, the device configurations and mechanical properties are important. The active electrode materials, including carbon-based materials, metal oxides, and hybrid composites, are detailed. Fiber-based supercapacitors facilitate fruitful textile integration, and three major types of fiber devices like parallel, twist, and coaxial structures are outlined. Stretchable fiber supercapacitors offer stable performances under large-level deformation in all directions and also can be twisted and stretched. Woven fiber supercapacitors into a large area of fabric substrate realize self-powered wearable devices. The innovative materials, including intrinsically stretchable and structurally stretchable materials, are key driving factors in the development of stretchable supercapacitors. Architecture designs like wavy, buckled, and island-bridge patterns are crucial for fabricating stretchable supercapacitors that can be applied on wearable bioelectronics and implantable bioelectronics as power supplies and energy storage units. Additional features such as biocompatibility with tissues, biodegradability, and adaptivity while maintaining electrochemical performance are crucial for all wearable supercapacitors. When integrated with bioelectronics system for healthcare monitoring, the wearable supercapacitors can accelerate the advances in implantable biointegrated devices and smooth their way into clinical applications. The developments in these hybrid multifunctional systems will enable next-generation diagnostic and therapeutic devices that contribute to public health.

3 Flexible and stretchable batteries

3.1 Introduction

In batteries, the electrical energy is stored indirectly as potentially available chemical energy. The electrochemically active molecules in batteries undergo Faradaic oxidation or reduction reactions and liberate charges that flow between electrodes having different potentials. Batteries are available for single use or multiple uses, and therefore, the battery systems can be classified as primary and secondary batteries (239). The primary batteries use the active chemical species only once and after a discharge they need to be thrown out. But the secondary batteries can be charged by applying electricity and the charging process enables conversion of the active materials back into their initial state, and allow them for further discharge (240). While storing the electrochemical energy in batteries, chemical interconversions of the electrode materials happen along with phase changes. Even though the net energy change is happening in a reversible thermodynamic manner, the interconversions of the active electrode materials are often irreversible during the charge and discharge processes. Hence the battery cycle life is limited to several thousands of cycles (241).

Typically, a battery consists of a positive and a negative electrode, separated by an electrolyte which allows ionic conduction between the electrodes. Ideally the electrolyte should be an electronic insulator in order to avoid the possible self-discharge and the internal short-circuit issues (242). Ionic conductors such as concentrated acids, alkalis, salts, organic salt solutions, ceramics, polymers, and fused salts are commonly used as electrolytes. The electrodes should be placed as close to each other in order to eliminate the internal resistance of the cell, so that the voltage drop across the battery can be minimized. Usually the electrodes are separated by a thin, porous, insulating film where the pores are filled with the electrolyte, and ionic conduction happens between the electrodes. As the batteries are usually operating in the discharge mode, the positive electrode is termed as cathode and the negative electrode is known as anode. Typical metals used as negative active components are zinc (Zn), lead (Pb), cadmium (Cd), and lithium (Li). Metal oxides at higher oxidation state, metal sulfide, elemental oxygen, or bromine are generally used as positive active materials. In order to avoid the poor electronic conductivity of these materials, a conducting filler is usually mixed to promote the electrons to the reaction sites (241).

One of the best available rechargeable batteries is the lead-acid battery, which is widely used in automotive industry to supply high current needed for starter motors. This low-cost battery uses lead dioxide as the positive electrode material and lead as the negative active-species, and sulfuric acid as the electrolyte. The poor weight-to-energy density ratio of the lead acid battery restricts its application for

https://doi.org/10.1515/9781501521287-003

wearable electronics application (243). Lithium-ion batteries (LIBs) have been explored as one of the most predominant types of power sources because of their advanced characteristics such as high energy density, long cycle life, no memory effect, and wide working potential. But the majority of commercialized LIBs – coin-type, cylindrical, or prismatic – are rigid and not suitable for wearable application. The conformal integration of batteries with the epidermis or other textile substrates requires appreciable flexibility and stretchability to sustain the performances during irregular body motion or muscle movements (244). Hence it is significant to develop bendable, stretchable, and portable LIBs with high energy densities.

Lithium possesses the lowest mass density among metals, most negative reduction potential (–3.04 V versus standard hydrogen electrode), and high theoretical capacity (3,860 mAh g^{-1}); hence, it is considered as one of the most promising anode materials (245). During discharging of LIBs, Li$^+$ ions are detached from the anode surface and transferred through the electrolyte and inserted into the cathode, and electrons move from the anode to cathode via the external circuit (246). This process is reversed during the charging process, and the reversible transfer of Li$^+$ ions between the electrodes realizes energy delivery and storage in the LIBs. Some bottlenecks still exist for LIBs which are the instability of Li metal and the formation of lithium dendrites (247–249). Li metal needs to be handled properly to avoid major safety problems like fire or even explosion; this is potentially important for all wearable applications. The growth of lithium dendrites in LIBs causes low coulombic efficiency and serious safety hazards. Owing to their ultra low mass density and high energy density, immense research is focused on to develop flexible and stretchable LIBs without having any safety issues (250).

Beyond lithium, batteries based on lithium metal anode and other cathode materials such as sulfur (Li-S) and oxygen (Li-O$_2$) possess higher theoretical capacities than the normal Li-ion-based batteries (251–253). In order to address the safety problems, researchers are exploring other materials like Zn, Mg, Na, and Al due to their low cost and high abundance (254). The volumetric/gravimetric capacities of these materials offer promising alternative platforms for lithium (255). This chapter discusses the developments of different architectures of flexible and stretchable batteries, mostly based on LIBs, and other latest developments based on various battery systems.

3.2 Fiber-based flexible batteries

Fiber-based textiles are soft, deformable, flexible, breathable, and durable, and thus they are an ideal substrate for wearable application. A fiber-shaped battery offers several advantages than the conventional planar configuration. The fiber battery can maintain its performances under repeated bending, twisting, and even stretching (256). The fiber-shaped electrodes can be woven into cloths with porous structures that enable transport of air and water vapors which are much needed for wearable textiles. The fibers can be seamlessly attached to substrates having variety of curved surfaces. The fiber battery can be constructed by twisting two functionalized fiber electrodes (257). The planar batteries possess major components like active material, conductive filler, binder, separator, and current collector. Similarly, the fiber-based batteries should configure vigilantly to establish firmly anchored active materials on conductive substrate with good electrical conductivities and mechanical resiliencies (258).

3.2.1 Fiber-based flexible LIBs

Materials that are typically utilized as anode materials for LIBs are metallic lith-ium (259), graphite (260), hard carbon (261), lithium-based perovskites (262), sil-icon oxides (263, 264), and tin-based alloys (265). Conventionally used cathode materials are lithium cobalt oxide (266), lithium manganese oxide (267), lithium nickel cobalt manganese oxide (268), lithium ion phosphate ($LiFePO_4$) (269), FeS_2 (270), V_2O_5 (271), and so on. Ion-conducting materials such as $LiPF_6$ (272), $LiClO_4$ (273), $LiAsF_6$ (274), and $LiCF_3SO_3$ (275) are usually used as electrolytes (276). Typically non-aqueous solvents are used for both primary and secondary LIBs as they have wide window of electrochemical stability that exceeds 4 V. The commonly used organic solvents include cyclic esters like ethylene carbonate, propyl-ene carbonate, γ-butyrolactone, and cyclic ethers like 1–3 dioxolane, 2-methyltetrahy-drofuran, and linear ether like 1,2-dimethoxyethane. Mixtures of solvents are usually preferred in order to achieve high ionic conductivity and stability of anode materials in LIBs (277). Many of the organic electrolytes are flammable and cause major safety problems when batteries are exposed to moisture or high temperature, overcharged or short-circuited internally (278). Aqueous recharge-able LIBs are promising systems owing to their safety, environmental friendli-ness, and low capital investment. Moreover, the organic electrolyte solution can be replaced by greener, safer, and non-flammable aqueous electrolytes having good ionic conductivities. The aqueous electrolytes possess ionic conductivities of about 2 orders of magnitude higher than the organic electrolytes that provide high rate capacity and low overpotential. But due to the high electropositive na-ture of lithium metal, decomposition of water occurs rapidly and thus aqueous electrolytes are suitable for low power requirements.

Robust incorporation of the active materials on the conductive fibrous material is essential for realizing flexible and light weight fiber LIBs. The aligned CNT fibers are good support materials as they promote stable anchoring of active materials to enable reversible intercalation and deintercalation of Li^+ ions. CNTs offer high sur-face area and good electronic conductivity and thus they can act as both structural support and current collector for charge transport. As the use of binder or external metallic current collector is not required, the fiber-based LIBs provide high specific capacities (279). Major techniques available for preparing CNT fibers are wet spinning (280), array drawing (281), and gas phase spinning (282). The wet spinning involves extruding the polymer solution through a spinneret into a solvent–non-solvent mixture. The mutual diffusion of polymer solution and the mixture coagulates to form fibers. The major drawback of wet spinning method to produce CNT fibers is it necessitates time-consuming pretreatment processes. The array drawing method involves direct spinning of fibers from vertically aligned CNT arrays by exploiting the van der Waals force between the CNTs. But the fiber

Table 3.1: Various LIBs constructed by polymer, metal, and CNT-based fibers, modified with different electrode materials, and their properties.

Material (Anode/Cathode)	Fiber diameter	Structural configuration	Conductivity/ resistance	Capacity	References
Polyester@CNT/ $Li_4Ti_5O_{12}$ Polyester@CNT/ $LiFePO_4$	N/A	3D textile structure	1,300 S cm^{-1}	454 mAh cm^{-3} (162 mAh g^{-1})	(290)
Nylon@Ag/Zn Nylon@Ag/MnO_2	N/A	Stretchable	0.59 Ω cm^{-1}	7.75 mAh	(291)
Cotton@CNT/Si Cotton@CNT/ $LiMn_2O_4$	2 mm	Coaxial	N/A	0.44 mAh (2 cm)	(279)
Cu wire@Ni-Sn Al wire@$LiCoO_2$	1.2 mm	Cable	N/A	25 mAh (25 cm)	(153)
AgNW/PDMS@Zn AgNW/PDMS@Ag	N/A	Stretchable twisted	9.8 Ω	0.011 mAh (0.1 cm2)	(292)
Cu wire@Zn Stainless steel@Ag	2.5 mm	Wire structure	7 to 15 Ω	12.2 mAh (10 cm)	(293)
CNT@$Li_4Ti_5O_{12}$ CNT@LiMn2O_4	160 and 130 µm	Twisted Stretchable	N/A	0.36 mAh (1 m)	(294)
Li wire CNT	N/A	Coaxial	N/A	37.5 mAh (15 cm)	(295)
CNT@$Li_4Ti_5O_{12}$ CNT@$LiMn_2O_4$	130 and 70 µm	Twisted	N/A	0.016 mAh (10 cm)	(296)
Li wire CNT@Si	30 to 60 µm	Twisted	N/A	200.4 mAh cm^{-3} (1,670 mAh g^{-1})	(297)
Li wire CNT@MnO_2	2 to 30 µm	Twisted	103 S cm^{-1}	109.6 mAh cm^{-3} (218.3 mAh g^{-1})	(298)

length is really limited to the size of the CNT arrays. Windle et al. reported direct spinning of CNTs in the gas phase from an aerogel which is suitable for large-scale production of CNT fibers with high mechanical strength and electrical conductivity (283). Liang et al. reported a CNT fiber-based aqueous LIBs, where the fibers are prepared by directly spinning through a floating catalyst chemical vapor deposition method and then functionalized with $LiFePO_4$ and lithium titanium phosphate ($LiTi_2(PO_4)_3$) as cathode and anode materials, respectively (284). The composite fiber electrode assembled LIB showed high specific capacity of 29.1 mAh g^{-1} at a current density of 0.25 A g^{-1}. The battery showed stable specific capacity under various degrees of bending deformations which are essential for flexible energy storage applications.

Silicon is a well-studied electrochemically active anode material due to its highest theoretical specific capacity. But it suffers from a large change in volume during the lithiation/delithiation process and causes hasty capacity decay. The drawback can be nullified by introducing a second phase with interfacial nanostructures that can accommodate the volume change. Usually, voids are created between the two phases (285, 286) or formation of scaffolds at the second phase (287, 288). The excellent structure and mechanical properties of CNTs are appropriate to function as the second phase and the silicon deposition on CNT arrays possess good electrochemical performances and charge transport kinetics. This strategy can be utilized for fiber-shaped coaxial LIBs by winding two CNT composite yarns with a cotton fiber (289). Peng et al. developed such aligned CNT yarns coated with silicon that can functions as anode and accommodates high volume change of silicon. As lithium manganate possesses large working voltage, safety, low-cost, and structural stability, it can be incorporated with CNT yarn to function as cathode. Such coaxial fiber LIB showed good linear capacity of 0.22 mAh cm^{-1} and, when the fiber is woven into a textile, it shows an areal energy density of 4.5 mWh cm^{-2}. Apart from CNT fibers, metal- and polymer-based fibers are extensively utilized for developing flexible high-performance LIBs. Table 3.1 summarizes various electrode materials, structural properties, and capacities of different fiber-based LIBs (152, 279, 290–298). These flexible fiber batteries offer weavability and miniaturization possibilities that make them appropriate for realizing fabric-based wearable batteries.

3.2.2 Fiber-based flexible metal-air batteries

Metal-air batteries have been extensively studied in the past decade which can generate electricity via a redox reaction between metal and oxygen present in air medium. The open cell feature of these batteries admits the free availability of oxygen as active cathode material. The oxygen consumption of these batteries resembles the functioning of a fuel cell, and they usually provide much higher energy densities than LIBs. The $Li\text{-}O_2$ batteries possess much higher theoretical energy density

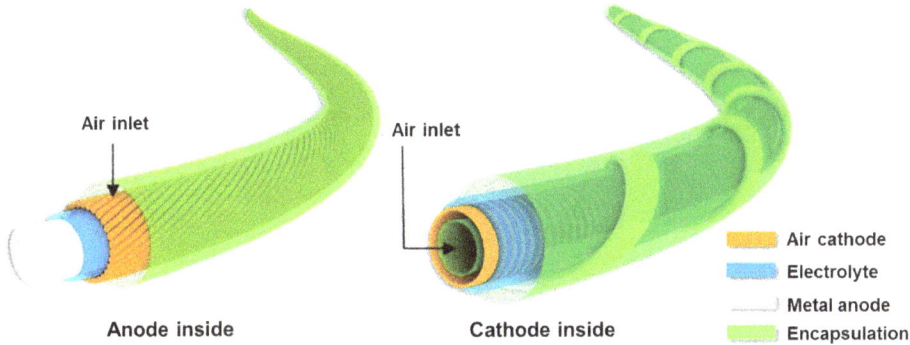

Figure 3.1: Schematic illustration of two types of fiber-shaped metal-air battery architectures (reproduced with permission from ref. (299)).

(3,600 Wh kg^{-1}) than LIBs (460 Wh kg^{-1}). The fiber-shaped metal-air battery facilitates 360° solid–liquid–air interface that is particularly useful to consume maximum oxygen from air. Apart from Li-air batteries, Zn-air, Al-air, and Li-CO$_2$ batteries are also recently advanced to provide greater power storage capacity (251).

The recent progresses involve two different configurations of coaxial architectures for 1D metal-air batteries (Figure 3.1). These assembly schemes can be named as anode inside and cathode-inside batteries (299). In the anode inside type, the 1D metal anode is first created, and then wrapped by an air cathode membrane separated with a gel electrolyte. This configuration possesses several benefits: (i) the inner metal anode is protected with multiple outer layers, thus efficiently preventing its corrosion by atmospheric moisture or gas; (ii) the outer air cathode with large diameter provides absolute exposure to the surrounding air and thus augmented current density. In the case of cathode-inside metal-air batteries, air cathodes are fabricated in a central ventilation manner to realize air inlet, and the flexible metal wire/strips coated with electrolyte are then wrapped around as the anode outside [45]. This prototype is promising for metal-air batteries that require pure oxygen as the source, which can be directly pumped from the gas container into tubular air cathodes.

When metal-air batteries are used with aqueous electrolytes, the underlying discharge mechanisms are described as equation (3.1) (M = Zn/Mg/Al/Fe). At anode, the released metal ions react with hydroxyl radicals from aqueous electrolytes form $M(OH)_n$, and the reduction of environmental O$_2$ is happening at the cathode region.

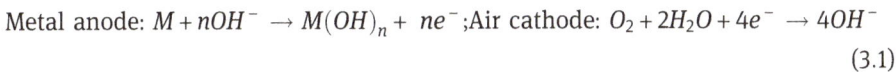

Metal anode: $M + nOH^- \rightarrow M(OH)_n + ne^-$;Air cathode: $O_2 + 2H_2O + 4e^- \rightarrow 4OH^-$

$$(3.1)$$

In the case of metal-air batteries with aprotic electrolytes, the reaction proceeds as showed in equation (3.2) (M = Li/Na/K). On metal-based anodes, the reversible adsorption/stripping process enables the charge/discharge reactions. But in cathodes, the oxygen reduction or oxygen evolution reaction causes the reversible generation of superoxide or peroxide. In Li-CO_2 batteries, CO_2 is utilized as the cathode reagent, and Li_2CO_3 and C are formed as the discharge products (300).

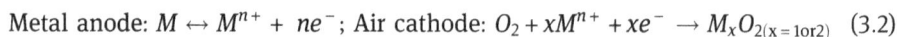

$$\text{Metal anode: } M \leftrightarrow M^{n+} + ne^-; \text{ Air cathode: } O_2 + xM^{n+} + xe^- \rightarrow M_xO_{2(x=1or2)} \quad (3.2)$$

3.2.2.1 Li-O_2 batteries

A non-aqueous Li-O_2 battery typically consists of three critical components such as a lithium anode, a carbon-based porous cathode mostly with catalysts, and Li^+-containing non-aqueous electrolyte. The crucial charge/discharge reactions are happening at oxygen electrode, and during discharging, O_2 receives electrons and combine with Li^+ from the electrolyte to form Li_2O_2 at the cathode. During charging, the reverse reaction occurs which is accompanied by huge polarizations with a voltage gap of 1.2 V. The polarization leads to the decomposition of electrolytes that promotes an insulating coating over the electrodes which further enhances the overpotential and accelerates the decay of electrolytes. The sluggish electron transfer rates further leads to poor rate capability, coulombic efficiency, and short cycle life. Utilizing efficient electrocatalysts for improving oxygen reduction or evolution reactions would enhance the performances of the Li-O_2 batteries. Researchers used many catalysts like porous carbon (301), noble metals (302), transition metal oxides (303), perovskite compounds (304), and nitrides (305). Besides, soluble redox mediators are also explored (306, 307) for catalyzing O_2 reduction and augment the level of O_2^- in solution that could result in an exceptional improvement of discharge capacity (306). A highly conductive porous configuration facilitates a large specific surface area, and a high rate of electron and oxygen transportations. The coaxial structure usually maximizes the effective interfacial area for gas adsorption in comparison with other structures of fiber-based batteries. The vertically aligned coral-like nitrogen-doped carbon nanofiber interfaced by microporous stainless steel cloth is shown to be useful for developing reversible Li-O_2 batteries with a narrow voltage gap of 0.3 V between charge/discharge plateaus (308).

3.2.2.2 Zn-air batteries

The zinc-air batteries are inexpensive and environmentally friendly, and possess high theoretical energy density of 1,086 Wh kg^{-1}; hence, they are suitable for powering wearable electronic devices. Similar to Li-O_2 batteries, the zinc-air battery necessitates efficient electrocatalysts for promoting discharge/charge processes. In order to overcome the reluctant electron transfer kinetics, the fiber electrodes must functionalize with porous conductive catalyst having favorable mechanical properties. The benchmark catalysts used for high-performance oxygen reduction or evolution reactions are Pt-, Ir-, or Ru-based compounds (309–311). In order to reduce the cost, inexpensive elements are widely used to replace the precious metal-based systems, including carbon, sulfides, and transition metal oxides (312, 313). Even though numerous bimetallic or trimetallic catalysts are developed, only a few of them are bifunctional which are able to catalyze both oxygen reduction and evolution reactions. Such catalysts can reduce the cost and simplify the manufacturing processes. The activity and performances of these nanostructured catalysts can be increased by merging with a conductive substrate. But the poor mechanical bonding with the substrate or the use of insulating polymer binder may reduce the activity of these catalysts and subsequently affect the performances of the batteries. This problem can be addressed by constructing a binder-free three-dimensional porous electrode configuration using bifunctional electrocatalysts.

Co_4N shows exceptional O_2 evolution activity because of its exceptional intrinsic electrical conductivity, but Co-based semiconducting oxides possess poor O_2 reduction abilities. Introducing transition metal-N co-ordination moieties with metallic Co_4N could effectively increase the O_2 reducing activities. This is a good method for developing bifunctional electrocatalysts for 3D freestanding Zn-air battery. Zhang et al. prepared a flexible bifunctional electrode consisting of carbon fibers and network of polypyrrole nanofibers embedded on carbon cloths (314). ZIF-67, a metal-organic frame work, is pyrolyzed on the polypyrrole fibers, wherein the nitrogenous gases evolved during pyrolysis are exploited to induce in situ conversion of ZIF-67 to Co_4N. The methodology eliminates the use of extra NH_3 for treating Co-based oxides/hydroxides, and the intertwined N-C fibers form a conductive 3D interconnected network. Additionally, the pearl-like structure of polypyrrole nanofibers facilitates to anchor Co_4N units and Co-N-C active sites on to the N-C fibers. These functional electrodes exhibit excellent electrocatalytic properties like reduced overpotential (310 mV at 10 mA cm^{-2}) for O_2 evolution and a half-wave potential of 0.8 V for O_2 reduction reactions. Also, the flexible Zn-air batteries showed a low discharge–charge voltage gap (1.09 V at 50 mA cm^{-2}), a stable retention in current density (20 h), and long cycle life of 408 cycles.

3.2.2.3 Al-air batteries

Aluminum-air batteries are a very promising technology for powering automotive vehicles and wearable electronics. Al-air battery possesses the theoretical specific energy density of 8.1 kWh kg^{-1}, which is around ten times higher than that of currently available Li batteries. Aluminum is the third most abundant element in the earth crest that enables low-cost production of high-performance batteries. Al-air batteries involve simple fabrication process based on an aqueous electrolyte, aluminum anode and an air cathode. The reaction mechanisms associated with the anodic and cathodic half cell, when an alkaline electrolyte is employed, can be described by the following equations (3.3) and (3.4), respectively:

Anode: $Al + 3OH^- \rightarrow Al(OH)_3 + 3e^-$ (*at pH* 14, *potential* -2.31 *V*)   (3.3)

Cathode: $O_2 + 2H_2O + 4e^- \rightarrow 4OH^-$ (*at pH* 14, *potential* $+0.40$ *V*)   (3.4)

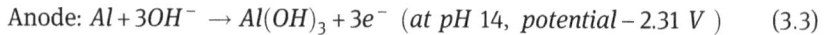

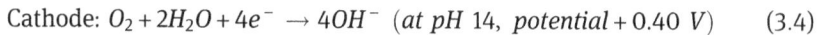

The net reaction possesses the theoretical open-circuit voltage of 2.71 V (equation (3.5)):

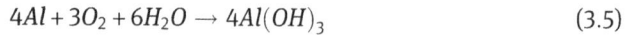

$$4Al + 3O_2 + 6H_2O \rightarrow 4Al(OH)_3 \tag{3.5}$$

A major drawback involved in Al-air batteries is the corrosion of aluminum and its surface passivation by formation of a protecting oxide coating. The corrosion is associated with evolution of hydrogen as shown in equation (3.6):

$$2Al + 6H_2O \rightarrow 2Al(OH)_3 + 3H_2 \tag{3.6}$$

The corrosion issue could be resolved by using specific aluminum alloys as anode material or by using solution phase corrosion inhibitors (315, 316). Researchers are also parallelly working to exploit the H_2 evolution as an effective energy source, thus optimizing the parameters required for H_2 production, its collection, and its productive storage methods.

Fiber-based Al-air batteries can be prepared by using a spring like Al substrate. A gel electrolyte was sequentially coated and wrapped by hybrid sheets of cross-stacked CNT-silver nanoparticles (317). In this design, the aligned CNT sheets provide a porous network to productively absorb oxygen, and the embedded nanoparticles function as an efficient catalyst so as to enhance the energy densities. The alkaline hydrogel electrolyte consists of poly(vinyl alcohol) and poly(ethylene oxide) with ZnO and Na_2SnO_3 as additives that eliminate the corrosion of aluminum in alkaline medium. The stretchable nature of Al spring, stacked CNT sheets, and hydrogel electrolyte enables to construct flexible and stretchable Al-air batteries. These fiber-shaped batteries could easily be woven into textile substrates to construct self-powered wearable electronics.

The nitrogen-doped carbonaceous materials are potential candidates for functioning as air cathodes in metal-air batteries. In particular, transition metals incorporated with carbon-based materials when doped with nitrogen showed exceptional O_2 reduction activities similar to that of Pt-based catalysts. The transition metal carbides like Fe_3C showed increased tendency to bind nitrogen, thus forming N_xC and FeN_x as active catalytic sites. Wang et al. developed a distinctive host–guest model for air cathode in solid-state Al-air battery (318). Fe_3C nanophases encapsulated in N-doped carbon nanofiber was synthesized by electrospinning method and used as catalysts. The use of Fe-based metal organic frameworks facilitates enhanced specific surface area, nanocrystallinity, mesoporosity, and the electrospinning method provides cross-linked porous network configurations. The active sites of Fe3C nanoparticles embedded in the carbon shell provide good interaction between the catalyst and the nanofibers. The mesoporous structure increases O_2 uptake capabilities, and the N-doping enhances the electrical conductivities. Such 3D network configuration with high-performance O_2 reduction activities is essential for developing mechanically flexible fiber-based batteries.

3.2.2.4 Li-CO₂ batteries

Metal-CO_2 batteries offer value added CO_2 conversion by electrochemical CO_2 splitting with surplus electricity. Metal-CO_2 batteries originated from the analysis of metal-O_2 batteries in atmospheric air circumstances. In the presence of air, metal-air batteries showed reduced stability and performance when compared to those working in pure O_2 due to the formation of carbonate from CO_2. But studies revealed that non-aqueous Li-O_2 batteries possess a threefold augmented capacity when purged with a certain quantity of CO_2 along with the O_2 gas (319). Lithium undergoes electrochemical reaction with CO_2 in presence of O_2 forming Li_2CO_3 which is more chemically stable than Li_2O_2 (equation (3.7)).

$$2Li + \frac{1}{2}O_2 + CO_2 \rightarrow Li_2CO_3 \tag{3.7}$$

Because of the thermodynamically favorable driving forces, the initially formed Li_2O_2 in Li-O_2 batteries can be easily converted into Li_2CO_3 in presence of CO_2. CO_2 gas possesses high solubility in organic electrolytes, which is 50 times higher than O_2 gas. This facilitates a high degree of CO_2 utilization even though the CO_2 content is limited in air (0.03 vol%).

The major challenges for realizing 1D flexible Li-CO_2 batteries lie in proper designing and fabrication of fiber-shaped cathodes with enormous catalytic properties and mechanical deformability. It is obvious that the coating of active catalyst using a polymeric binder on fiber-shaped current collector is not suitable as it suffer from poor flexibility. In addition, introducing the polymer binder masks the carbon nanophases, limits the catalytic activity, and causes parasitic reactions. Hence, developing conductive flexible fiber electrodes with homogenously distributed high-performance catalysts is much needed for flexible Li-CO_2 batteries. Wang et al. developed a binder-free fiber-based freestanding cathode for Li-CO_2 batteries by introducing N-doped surface-wrinkled CNT arrays homogenously attached on Ti wires (320). The high conductivity of metal wires and the presence of more defects and active sites typically promote both CO_2 reduction and evolution reactions. The fiber battery provides excellent mechanical resiliency, high discharge capacity of 9292.3 mAh g^{-1}, good rate capability, and cycling stability for 45 cycles.

3.2.3 Fiber-based metal-sulfur batteries

Lithium ion batteries offer the highest energy density that an intercalation material can provide (300 mA h g^{-1}). New systems are being developed to further exceed the energy density and reduce the manufacturing cost. The next-generation batteries are moving beyond intercalation materials but exploiting integration chemistry–based materials where covalent bonds undergo cleavage/formation along with structural or morphological changes during discharge/charge cycles. Sulfur possesses several significant characteristics like low equivalent weight, non-toxicity, and low cost. A range of promising batteries are constructed with sulfur-based cathodes and alkali metal anodes such as Li, Na, K, Mg, Ca, and Al.

3.2.3.1 Li-S batteries

Li-S batteries offer much higher energy density at a lower cost compared with LIBs. The theoretical energy density can reach beyond 2,500 W h kg^{-1} assuming absolute reaction to form Li$_2$S. In an Li-S battery, Li$^+$/Li serves as the overall redox couple at an average voltage of 2.15 V as per the equation (3.8) (321):

$$S_8 + 16Li^+ + 16e^- \leftrightarrow 8Li_2S \qquad (3.8)$$

Some significant challenges exist for Li-S cell which must be considered while fabricating fiber-based Li-S batteries. The insulating behavior of sulfur and its discharge products mandate substantial integration and linkages with the conductive current collectors. The solubility of long-chain polysulfide ions formed during reduction of S$_8$ causes high mobility in solution phase to the lithium anode via the separator, where they reduce into insoluble Li$_2$S$_2$ or Li$_2$S. The complete coverage of S$_n^{2-}$ over the Li anode enables reaction with fully reduced sulfides yielding lower order polysulfides which further diffuse back to the cathode and then re-oxidize into the polysulfide ions. Such parasitic reaction causes internal shuttling, reduced utilization of active mass during the discharge process, and diminished coulombic efficiency during the charging process.

Researchers found that the introduction of lithium nitrate as an additive can decreases the shuttle phenomena for Li-S batteries. The elimination of shuttling increases the coulombic efficiency of the redox reaction. Porous conductive carbon materials like activated carbon, mesoporous carbon, carbon fiber, and CNTs augment the surface area that makes them suitable substrate for sulfur cathodes. Also the encapsulation of sulfur in carbon-based materials facilitates functioning as an active mass and irradiate the diffusion possibilities and the shuttling issues. Aurbach et al. utilized microporous activated carbon fibers having elemental molten sulfur for preparing binder-free sulfur cathodes. The use of carbon cloth enables high degrees of flexibility and allows high loading of sulfur without affecting its conductivity and excellent electrochemical performances (322). Wu et al. developed a carbon fiber/sulfur composite electrode as cathode for Li-S cell using electrospinning approach and nickel acetate as a catalyst for graphitization at lower temperature (323). The carbon fiber having hollow graphitized carbon spheres enables encapsulating sulfur into the cage-like structures.

The configuration prevents the dissolution of polysulfides, and provides good electrical conductivities.

3.2.3.2 Na-S batteries

The Na-S battery possesses many significant advantages such as high theoretical specific energy density of 760 W h kg^{-1}, low cost (both sulfur and sodium are abundant in nature), low self-discharge rate, and high power density. Conventionally, Na-S batteries are operated at elevated temperatures (300–350 °C) with molten active materials separated by a solid-state ceramic electrolyte. Later Na-S batteries are advanced to operate at room temperature by exploiting liquid organic electrolytes with high Na ion conductivity where Na-S redox reactions happen at room temperature. In room temperature Na-S batteries, Na metal functions as anode, sulfur composite electrode as cathode, and an organic solvent with dissolved Na salt as electrolyte along with a separator. Choi et al. utilized one-dimensional fibrous configuration for constructing flexible Na-S batteries by electrospinning of polyacrylonitrile nanofibers (324). The presence of sulfur–carbon bonds and sulfur nanodomains in the conductive carbon platform realizes robust performance during repeated charge–discharge redox cycles. The nanofibrous morphology provides high interfacial surface area for effective electrode reactions and thus enables high rate capability.

3.2.3.3 K-S batteries

Potassium-sulfur batteries are recently attracting substantial attention due to their electrochemical advantages. The K-S battery possesses a theoretical gravimetric energy density of 1,023 Wh kg^{-1} for an average potential of 2.1 V. Potassium possesses lower redox potential than Na and Li that implies the high voltage window for full cell devices. Potassium is a weaker Lewis acid compared to Na and Li and possesses smaller Stokes' radius of 3.6 Å in carbonate solvents. This facilitates increased ion mobility, ion transport number, and ionic conductivity for K$^+$ ions. Some challenges still exist for potassium-based batteries that could affect the real performances. The larger ionic radius of K$^+$ (0.138 nm) limits its solid-state diffusion in most ion active materials. The effective intercalation of K$^+$ into graphite necessitates slow reaction kinetics, and K metal anode is vulnerable to dendrites and to unstable solid electrolyte interphases (325). A flexible potassium ion battery is reported with promising anode materials which eliminate the possibility of volumetric expansion and structural disintegration during the charge-discharge cycles. The flexible sub-micron-sized carbon fibers wrapped in carbon nanotubes are prepared by electrospinning method, and the battery provides a capacity of 74.0 mAh g^{-1} at 20 mA g^{-1} (326).

3.2.3.4 Mg-S batteries

Rechargeable magnesium batteries are highly cost-effective and environmental-friendly energy storage systems. Magnesium is a highly abundant element present in earth crust which holds a large volumetric specific capacity (3,833 mAh cm^{-3}) and a low reduction potential of −2.37 V vs. standard hydrogen electrode. These multivalent metal ions are not compatible with majority of organic liquid electrolytes; hence, the metal ion transport is a critical challenge. Also the internal shuttle effects due to the formation of polysulfide intermediates needs to be resolved for long cycling performances. Although one-dimensional Mg-S batteries are not yet reported, studies are progressing on to realize flexible systems. Mitra et al. utilized carbon cloth as a flexible current collector which can overcome the sluggish electron transfer kinetics and avoids the dissolution of intermediate magnesium polysulfides. The 3D interconnected porous nature of carbon cloth facilitates high electrolyte accessibility into the pores and thus enhances the ionic conductivity. The inherent hydrophobic nature of the carbon cloth is not adequate for adsorbing magnesium polysulfide intermediates. The leaching issues can be addressed by inducing polarity to the carbon cloth with hetero-atoms. The N- and S-doped carbon cloth facilitates excellent chemical interaction with the polysulfides and thus maintains the cycling stability of Mg-S batteries (327).

3.2.3.5 Al-S batteries

Aluminum is the most abundant metal on the earth. Because of the ability of providing three electrons from an Al atom, it possesses large specific gravimetric capacity of 2,980 mAh g^{-1}. The high density of Al (2.7 g cm^{-3}) realizes high volumetric specific capacity 8,050 mAh cm^{-3} that makes Al a promising candidate for battery applications. Ionic liquids–based electrolytes are usually used for reversible coating/stripping of Al, and are completely safe as they cannot catch fire. The Al-S battery system typically follows the rocking-chair mechanism where Al^{3+} ions are intercalating into the cathode, without consuming any anions from the electrolyte. Ideally, the Al-S battery possesses a gravimetric energy density of 1,319 Wh kg^{-1} and a theoretical working voltage of 1.23 V. The cell reaction can be represented as equation (3.9) (328):

$$2Al + S_4^{2-} + 2OH^- + 4H_2O \rightarrow 2Al(OH)_3 + 4HS^- \tag{3.9}$$

Some research attempts have been reported to realize the Al-S batteries more flexible so as to withstand the mechanical strain. Manthiram et al. developed an Al-S battery based on flexible freestanding carbon nanofiber paper as cathode platform (329). A glass fiber coated with a layer of SWCNTs was used as the separator and an ionic liquid consisting of aluminum chloride and 1-ethyl-3-methylimidazolium chloride was utilized as electrolyte. The fiber paper offers a large interface structure and a robust conductive matrix to hold the active materials and the reaction intermediates/products. The SWCNT coating functions as an effective barrier to prevent the diffusion of reaction intermediates and as an upper current collector to diminish the polarization of the cell.

3.2.4 Fiber-based Ni-Fe batteries

Alkaline aqueous nickel-iron battery shows a theoretical capacity of 241.5 mA h g^{-1}. The battery possesses several merits like abundant active materials, being non-toxic and safe, but they suffer from lower energy and power densities compared to LIBs. In order to overcome the inactivation problem related to the Ni-Fe batteries and to improve their performances, an effective strategy is to anchor nickel- and iron-based nanostructures on a nanocarbon substrate with high electrical conductivity. Doping is an effective method to introduce defects in the optimized structure of semiconducting materials for increasing the electrocatalytic performances, rate capability, and durability. Xiao et al. reported a facile room temperature process for doping manganese into NiO and functionalizing Fe_2O_3 nanostructures on one-dimensional copper fiber structures (330). The incorporation of manganese provides optimized electronic and band structures and reduces the adverse properties of NiO and Fe_2O_3, thus boosting up the carrier transport efficiency. The assembled flexible fiber battery showed excellent energy and power densities, and are suitable for wearable energy storage applications.

3.3 Fiber-based stretchable LIBs

The flexible fiber-shaped batteries work well under bending and twisting deformations, but their performances are affected when stretched. Developing stretchable batteries is crucial for integrating with textile-based wearable electronic devices or other epidermal electronic skin for healthcare and fitness monitoring applications. The conventional Li foil or Li wire is rigid and brittle in nature, and difficult to accommodate strain during elastic deformation due to its irreversible elongation properties (331). As Li metal is poorly stretchable, developing a composite of Li electrode with a stretchable current collector is adequate for developing fiber-based LIBs. The infusion technique offers a versatile route to fabricate super-stretchy fibers directly from molten lithium with lithophilic medium. Li et al. developed a fiber-based stretchable battery by combining the flexibility and robustness of CNTs and the lithophilic nature of ZnO nanowire arrays (332). The fabrication process of the fiber-shaped Li-ZnO@CNT anode is schematically depicted in Figure 3.2a. This method facilitates the formation of micron-sized 1D Li metal anode, which can adapt well with the dynamic movements when winded spirally

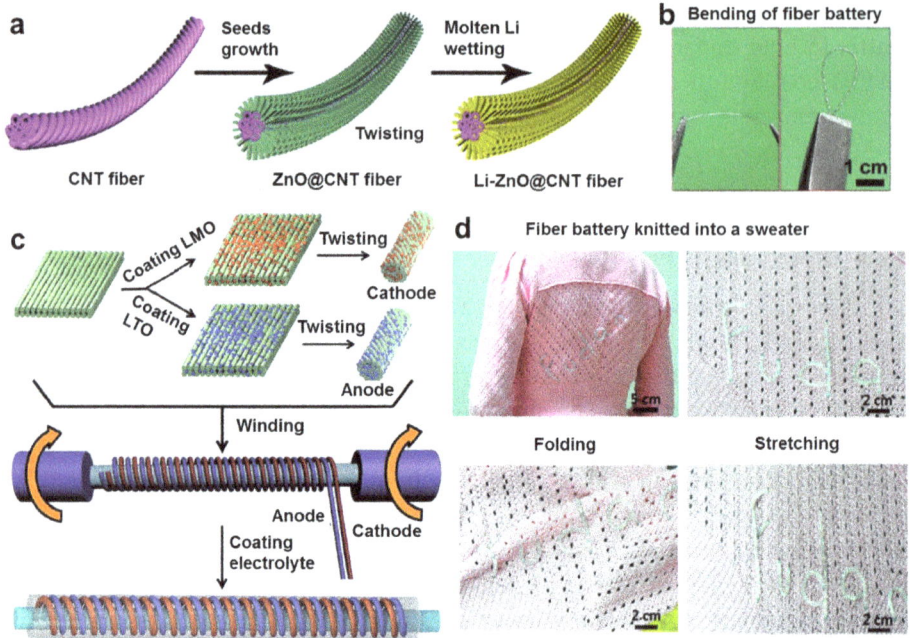

Figure 3.2: (a) Schematically representing the fabrication process of the fiber-shaped Li metal anode (Li-ZnO@CNT), and (b) shows the digital images of the fiber under straightening and bending states (reproduced with permission from ref. (332)). (c) Schematic demonstration of the preparation of super-stretchy fiber-based lithium-ion battery using MWCNT/LiMn$_2$O$_4$ and MWCNT/Li$_4$Ti$_5$O$_{12}$ as positive and negative electrode materials, respectively. (d) Digital images of stretchable fiber-shaped battery woven into a knitted sweater under folding and stretching conditions (reproduced with permission from ref. (333).

on an elastic fiber, and exhibits good bendability (Figure 3.2b). The three-dimensional ZnO nanowire function as an inducer for Li infusion and the CNT networks provides appreciable stretchability of 100%. The composite type electrode possesses dendrite-free morphology when cycled at various current densities and provides good electro-chemical performance with a limited capacity loss (4%) in 100 cycles.

Peng et al. reported another method to develop elastic fiber-based LIBs by winding functionalized lithium-based fiber electrodes around a stretchable PDMS fiber (333). Aligned CNT fibers are prepared and modified with LiMn$_2$O$_4$ and Li$_4$Ti$_5$O$_{12}$ to function as cathode and anode, respectively, and then twisted to form composite fibers (Figure 3.2c). These stacked fibers at a specific pitch distance are wounded around a PDMS fiber and then coated with a stretchable poly-(ethylene oxide)-based gel electrolyte having lithium salt of bis(trifluoromethane)sulfonamide, and succinoni-trile as a plasticizer. The elastic substrate and the winding structure provide stretch-ability to the fiber-shaped battery and the stretching deformation increases the pitch distances between the stacked active electrodes. The stretchable gel provides proper anchoring of the fibers and stability to the electrodes. This stretchable fiber LIB showed a specific capacity of 91.3 mA h g^{-1} and the performances are maintained under large stretching deformation of 600%. In contrast to stretchable planar batteries, these fiber-shaped batteries showed good scalability and can be effortlessly woven into textiles. Figure 3.2d shows optical images of the integrated battery on a knitted sweater. The battery textile possesses excellent adaptability to the regular body motion and can accommodates folding and stretching deformations, which is promising for wearable applications.

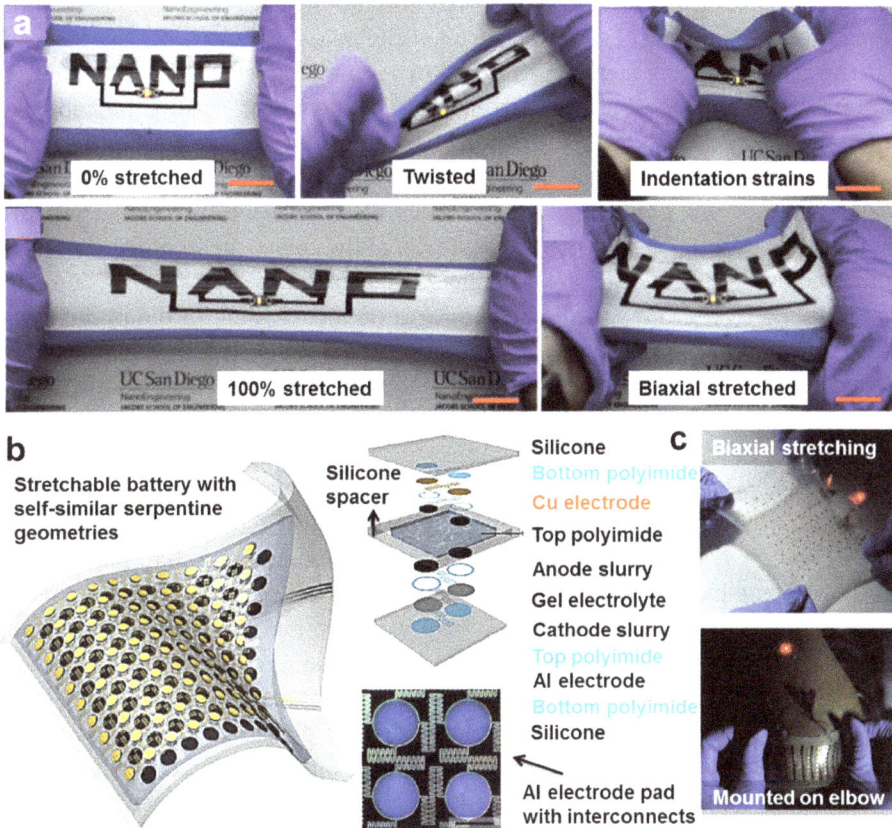

Figure 3.3: (a) Photographs of sealed Zn-Ag$_2$O battery when exposed to 0% stretching, twisting, indentation strains, 100% stretching, and biaxial stretching (scale Bar: 2.25 cm) (reproduced with permission from ref. (334)). (b) Schematic representation of a stretchable battery in a state of stretching and bending, and insets show the exploded view of the different layers in the battery and the optical image of the Al electrode pads having self-similar interconnects on a Si wafer.
(c) Operation of the stretchable battery connected to red light-emitting diode when biaxially stretched to 300%, and mounted on the human elbow (reproduced with permission from ref. (338)).

3.4 Intrinsically stretchable batteries

Stretchable batteries are generally prepared by following either random composite approach or deterministic composite methods. In random composite route, highly conductive fillers are percolated into an elastomeric matrix. The intrinsic stretchability of hyperelastic binders maintains electrical conductivity when the homogenously distributed fillers slide along each other during stretching. The conductive fillers like silver nanowires and CNTs provide robust electrical contact owing to their bundled percolation structures that keeps the network of connections via multiple pathways. Intrinsically stretchable batteries are realized by using screen, roll-to-roll, or inkjet printing of elastomeric composite inks.

The screen-printing technique allows large-scale production of high-resolution patterns by a simple cost-effective manner. The widely used coating methods like dip coating, spin coating, or spray coating necessitate low-viscosity composite inks. But screen printing needs a high viscous paste that enables large-scale loading of active materials and conductive fillers, which also realizes to precisely configure complex low-dimension patterns. The viscosity of the composite formulation can be optimized by components like binder, filler, active materials, and solvents. The elastic binder holds the components and controls the elastic limit of the printed patterns. The critical challenge associated with the random composite method is that the electrochemical performances of the active species and the elastic nature of the binders are mutually detrimental to the other.

Wang et al. developed a stretchable zinc–silver oxide rechargeable battery using screen printing of hyperelastic conductive composite inks (334). Zinc–silver oxide battery possesses a theoretical energy density of 300 W h kg^{-1} and provides the highest specific energy density of commonly available aqueous rechargeable batteries. Polystyrene-block-polyisoprene-blockpolystyrene (SIS) is one of the stretchable block copolymer consisting of long polyisoprene chain and short polystyrene terminal ends. SIS offers higher loading of active materials and maintains the mechanical and electrochemical performances of the printed traces. The as-developed Zn-Ag$_2$O battery showed a reversible capacity density of $\approx$2.5 mA h cm^{-2}, and the activity remains unaffected even after exposed to multiple 100% stretching iterations (Figure 3.3a). Recently, an all-printed, flexible, and rechargeable AgO-Zn battery was developed by layer-by-layer printing of stretchable composite inks to construct the current collectors, electrodes, and separators (335). The thermoplastic SEBS substrate-based battery provides ultra-high areal capacity (>54 mAh cm^{-2}), good rechargeability, and low impedance, and is useful for practical energy storage application for flexible electronics.

3.5 Serpentine interconnects-based stretchable batteries

The deterministic composite route offers engineered elasticity and also termed as is-land-bridge approach. In this method, the active rigid areas of stretchable batteries are interconnected by serpentine-shaped conductive bridges. The rigid islands are ulti-mately anchored onto the elastomeric substrate where the serpentine interconnects are freestanding. During stretching, the curved interconnects unwind and thus accommo-dates most of the strain and protects the active rigid islands. This method overcomes the limitations of the random composite route, where the performances of the active materials are diminished by the elastomeric binders. Lithographic methods are usually utilized for developing deterministic stretchable configuration which realizes high-reso-lution precise patterns with ultra-low dimensions (336). But the extensive applications of such devices are limited due to the non-compatibility of several active materials for lithographic fabrication. Merging of thin- and thick-film fabrication routes are suitable for addressing this problem. Wang et al. utilized screen-printing technique to fabricate thick film active islands and merge with the lithographically deposited thin ser-pentine interconnects (337). Such stretchable devices allow maximizing the loading of active components, and also the hybrid route allows preparing intrinsically stretch-able active islands.

Roger's group developed stretchable LIB that offers multiple characteristics like high storage capability, large-scale stretchability of 300%, and stable integration of wireless power transmission features to recharge the devices without direct physical contact (338). The advanced design with deformable self-similar interconnects offer high degrees of stretchability. The battery layout and design is schematically dem-onstrated in Figure 3.3b and the insets show the exploded view of the different layers in the battery configuration and the optical image of Al electrode pad with the interconnects. The battery supplies sufficient energy to power commercial light-emitting diodes having turn-on voltages of 1.7 V. The battery could be stretched up to 300% of its initial length, without affecting the intensity of the light-emitting diode, and found to be compliant when mounted onto the human skin (Figure 3.3c). The self-similar geometry involves iteratively applying curve-shaped serpentines be-ginning with a unit cell. Decreasing the dimension of the cell and interconnecting them in such a way to reproduce the original cell geometry is a feasible methodol-ogy to construct high orders of stretchable devices. Arias et al. developed a flexible and stretchable Ag-Zn battery with metallic current collectors having advanced me-chanical design like helical springs, spirals, and serpentines (339). By choosing the required geometry of current collectors, stretchable batteries can be prepared to in-tegrate with the wearable electronic devices. The use of helical band spring current collectors realizes flexible batteries with omnidirectional deformability, but the stretchable nature can be obtained by using serpentine ribbon geometry. The serpentine ribbons provide out-of-plane rotations, and the batteries can be oper-ated under 100% strain.

3.6 Kirigami-based stretchable batteries

A kirigami approach is another way to engineer elasticity in stretchable energy storage devices. Strain engineering is desirable to maintain the functional properties of the devices under external stress. As the stress failure causes the formation of micro-cracks, defects, and eventually poor electrical conductivity and device performance, the rigid and brittle materials must be engineered to manage the complexity of deformations. In order to increase the stretchability of conductive materials, the concept of art paper cutting in various traditions like kirigami, jianzhi, or silhouette can be adopted. The three-dimensional repetitive patterns on planar substrate possess considerable effects on withstanding mechanical deformations. The microscale kirigami patterning on nano-composite sheets provide stress delocalization over a number of deformation points. The kirigami method is also useful for predicting deformation mechanics for complex composite materials. By exploiting the standard photolithography techniques, micro-scale or nanoscale kirigami structures can be constructed, and the length scale can be controlled.

Jiang et al. developed a highly foldable LIB based on origami folding approach, where the planar battery is initially produced and then configured with foldable architectures (340). The method suffers from disadvantages like limited foldability from the folded to planar state, and difficulty in integrating folded patterns having

Table 3.2: Summary of the performances of recently reported stretchable batteries.

Battery type	Electrode layout	Stretch ability (%)	Output voltage (V)	Capacity	Energy density	Cycle retention	References
LIBs	Fiber	100	2.5	92.4 mA h g^{-1}	231 mWh g^{-1}	92.1% after 100 cycles at 0% strain	(342)
LIBs	Fiber	100	2.5	138 mAh g^{-1}	345 mWh g^{-1}	84% after 200 cycles at 100%	(343)
LIBs	Fiber	600	2.2	91.3 mA h g^{-1}	201 mWh g^{-1}	90% after 50 cycles at 100% strain	(333)
LIBs	Self-similar serpentine	300	2.2	1.1 mA h cm^{-2}	2.42 mWh cm^{-2}	~80% after 20 cycles at 0% strain	(338)
LIBs	Wavy layout	50	3.7	2.2 mA h cm^{-2}	8.14 mWh cm^{-2}	85% after 60 cycles at releasing state	(116)

Table 3.2 (continued)

Battery type	Electrode layout	Stretch ability (%)	Output voltage (V)	Capacity	Energy density	Cycle retention	References
LIBs	Wavy layout	450	2.4	1.1 mA h cm^{-2}	2.64 mWh cm^{-2}	~85% after 100 cycles at 450% strain	(344)
LIBs	Porous structure	80	1.75	120 mA h g^{-1}	210 mWh g^{-1}	70% after 300 cycles at 0% strain	(345)
LIBs	Origami	1300%	2.65	0.2 mA h cm^{-2}	0.53 mWh cm^{-2}	–	(340)
Zn-Ag	Coplanar	100	1.31	3 mA h cm^{-2}	3.93 mWh cm^{-2}	140% after 30 cycles at 100% strain	(334)
Zn-Ag	Coplanar	80	1.62	0.19 mA h cm^{-2}	0.31 mWh cm^{-2}	80% after 1,000 cycles at 80% strain	(346)
Zn-Ag	Serpentine	100	1.5	3.5 mA h cm^{-2}	5.25 mWh cm^{-2}	~100% after 20 cycles at 100% strain	(339)
Zn-MnO$_2$	Fiber	300	1.48	302 1 mA h g^{-1}	447 mWh g^{-1}	98.5% after 500 cycles at 0% strain	(347)
Zn-MnO$_2$	Coplanar	100	1.5	3.87 mA h cm^{-2}	5.80 mWh cm^{-2}	–	(348)
Zn-MnO$_2$	Coplanar	100	1.3	3.5 mA h cm^{-2}	4.55 mWh cm^{-2}	–	(349)
Na-ion battery	Porous structure	50	2.7	103 mA h g^{-1}	278 mWh g^{-1}	85% after 60 cycles at 50% strain	(350)
Al-air	Fiber	30	1.2	935 mA h g^{-1}	1,168 mWh g^{-1}	–	(317)
ALBs	Sandwich	100	1.2	100 mA h g^{-1}	120 mWh g^{-1}	80% after 15 cycles at 100% strain	(351)
ALBs	Coplanar	100	1.0	90 mA h g^{-1}	90 mWh g^{-1}	70% after 50 cycles at 100% strain	(352)
Li-air	Serpentine	100	2.7	7,111 mA h g^{-1}	2,540 Wh kg^{-1}	–	(353)

uneven surfaces with planar systems. But a kirigami method can combine both fold-ing and cutting procedures to merge patterns with even surfaces at the stretched state. This method provides high level of stretchability even at planar state and the mechanism is termed as plastic rolling. Jiang et al. utilized this method to prepare a stretchable LIB, based on slurry coated graphitic anode and $LiCoO_2$ as a cathode (341). After folding and cutting, the kirigami battery shows stretchability of over 150% and the performance is much similar to that of the planar device. The kiri-gami-based stretchable batteries have a large-scale scope for widespread applica-tion, including in the area of wearable electronics.

Stretchable batteries provide maximum mechanical flexibility and showed great importance for realizing emerging electronic devices like foldable and bendable smart phones or tablets. Table 3.2 shows the recent progresses involved in different types of stretchable batteries (116, 317, 333, 334, 338–340, 342–353). Attention must be focused onto the device design and the mechanical properties of all the device components such as binder, current collector, and gel electrolytes. Incorporating self-healing prop-erties to the electrode and the gel electrolytes is appreciable to minimize the damage and to sustain the performances.

3.7 Summary

The chapter discussed the advances in the wearable batteries for powering portable, skin-worn, or textile integrated devices. The representative examples of different types of wearable batteries and their architecture designs are detailed. Endowing the batteries with flexibility is necessary for compliant integration on irregular body surfaces and other wearable substrates. The fiber-shaped batteries provide several unique advantages like flexibility, weavability, and wearability. The flexible fiber batteries should possess remarkable electrical and mechanical properties without compromising their power and energy densities. Owing to lowest mass density of Li among metals, ultra-high theoretical capacity, and the most negative reduction potential, LIBs are the most promising for wearable applications. In addition LIBs possess extremely high theoretical energy density, but for practical applications, the dendrite growth must be suppressed and the coulombic efficiency should be improved by creating a stable interface. Stretchable batteries provide large-scale flexibility for maintaining the device performances under extreme strain. Design-induced stretchable batteries provide higher electrochemical performances over intrinsically stretchable systems due to the lack of insulative elastic binders. Flexi-ble and stretchable batteries represent a reliable platform for wearable energy storage application and by combining with energy harvesting devices; the hybrid system can provide stable continuous power output, particularly for healthcare monitoring devices.

4 Stretchable and self-healing gel electrolytes

4.1 Introduction

An electrolyte is a liquid or solid solution with the capability to conduct ions, but not conduction of electrons. Electrolytes are indispensable in all electrochemical energy storage devices, and their diversified chemistries make them significant for widespread applications (354). Depending on their physical state at room temperature, electrolytes are classified into liquid, solid, and gel type (355). The role of electrolytes in energy storage or conversion devices is to basically function as the medium required for charge transfer, in the form of ions, between the electrodes. Electrolytes consist of salts otherwise termed as "electrolyte solute" dissolved in aqueous or non-aqueous solvents. The aqueous electrolytes are more frequently used in supercapacitors but rarely in lithium-based batteries due to the stability problem with the electrodes (355, 356). Aqueous electrolytes are "green" electrolytes with very low cost, and provide higher ionic conductivity. The potential window for aqueous electrolytes is limited due to the decomposition of water, and the preferable operating voltage is around 1.23 V. However, the electrochemical window can be extended by eliminating the presence of O_2 and adjusting the pH values of the electrolyte. Although the term "non-aqueous solvent" has been extensively used in the literature, "aprotic" would be a more correct term. The active protons present in non-aqueous solvents such as ethanol or anhydrous ammonia cause stability issues with lithium. Typically, carbonate esters are used as organic solvents, and for LIBs, suitable lithium salts are combined to get enhanced performances. Factors like viscosity, dielectric constant, melting and boiling points, toxicity, flammability, and thermal stability must be considered in accordance with the practical applications. The organic solvents are compatible to the large electrochemical window and the current collectors, and show the ability to form robust solid–electrolyte interphases. The ionic conductivity and the interfacial properties of the electrolytes are crucial for energy storage application, and the performances are greatly affected by the physical states of electrolytes (357).

The liquid electrolytes possess high ionic conductivity and the capability to yield stable contact and surface wetting with electrodes. In order to obtain the complete advantage of the large specific surface area of nanostructured electrode materials, liquid electrolytes are more suitable. Also, liquid electrolytes attribute more wettability and accessibility toward porous electroactive nanomaterials (358). But there are some drawbacks for liquid electrolytes such as flammability, leakage, and instability. These issues are more critical for wearable application as majority of such devices are directly integrating on the skin surfaces. Gel-type electrolytes and solid polymer electrolytes are more suitable for constructing flexible energy storage devices owing to their leakage proof behavior, mechanical deformability and stability (359–363). But, the

https://doi.org/10.1515/9781501521287-004

poor interfacial bonding with the electrodes and the ionic conductivity compared with liquid electrolytes adversely affect their performances (364).

Introducing room temperature ionic liquids into the polymer electrolytes is suitable for addressing the safety issues owing to their non-flammability and high temperature stability (365). Additionally, ionic liquids show significant characteristics such as negligible vapor pressure and high electrochemical stability over a wide potential window (365–368). Table 4.1 shows the ionic conductivity and the electrochemical potential window of different gel electrolytes with ionic liquids (369–377). The nature and the constituents of electrolytes are different for supercapacitors and batteries, and the reaction mechanism and the operating environments are the crucial factors for designing the proper electrolytes. Wearable energy storage devices need to be flexible and stretchable in nature in order to mitigate the large strain developed during irregular body motion or muscle movements (378). Introducing electrolytes with similar mechanical properties to that of the flexible electrodes is essential for maintaining the solid interfacial contact and electrochemical performances (379). The stretchable electrolytes offer large degrees of interfacial bonding and avoid delamination from the electrode surface when exposed to stretching deformations (209, 380, 381). If the device is stretched beyond the elastic limit, the solid electrolyte layer undergoes crack formation which is detrimental to the device performances. In addition, if the electrolyte layer possesses the self-healing property, the autonomous repairing of microcracks could instantaneously regenerate the ionic conductivity and thus maintain the device performances (382–384).

4.2 Solid electrolytes

The liquid electrolytes offers high energy storage performances in all type of devices, but the growing demand for safety and engineering flexibility increases the demand for solid electrolytes. The next-generation supercapacitors and batteries need to be mechanically flexible to adapt with the three-dimensional curved surfaces, and should maintain their performances under bending, twisting, rolling, and stretching environments. Such multifunctional systems are suitable to integrate with biointegrated devices, wearable electronics, and microelectronics (385). Solid electrolyte is a promising platform to adapt with the mechanical deformations, and it allows the device performances under irregular movements or vibrations (386–388). Also, the high modulus of solid electrolytes minimizes the dendrite growth problems associated with lithium-metal batteries (389). Although solid electrolytes hold significant advantages, they face some drawbacks, including poor ionic conductivity, interfacial adhesion, and high interfacial resistance (390). Hence, solid electrolytes need to be formulated in such a way to minimize these issues and concomitantly balance the multifunctional abilities.

Table 4.1: Ionic liquids–based gel electrolytes and their properties.

Ionic gels	Ionic conductivity (S cm^{-1})	T (ºC)	Potential window (V)	References
PEO-PMMA ± LiN(CF$_3$SO$_2$)$_2$ ± TMHA[a]-LiN (CF$_3$SO$_2$)$_2$	10^{-5}	20	–	(369)
PEOEMA ± LiPF$_6$ ± BMI-PF$_6$	0.9 × 10^{-3}	25	4.30	(370)
PVDF-HFP ± LiN(CF$_3$SO$_2$)$_2$ ± BEP[b]-N (CF$_3$SO$_2$)$_2$	10^{-4}	20	5.50	(371)
PVDF-HFP ± P^{13}TFS[c] ± LiN(CF$_3$SO$_2$)$_2$	0.3 × 10^{-3}	20	5.75	(372)
PVDF-HFP ± Li(CF$_3$SO$_3$) ± DMOI[d]-CF$_3$SO$_2$	3.2 × 10^{-5}	20	5.00	(373)
PVDF-HFP ± LiPF$_6$ ± EMI[e]-PF$_6$	1.6 × 10^{-3}	20	5.00	(374)
PVDF-HFP ± LiN(C$_2$F$_5$SO$_2$)$_2$ ± EMI-N (C$_2$F$_5$SO$_2$)$_2$	10^{-3}	25	5.50	(375)
P(EO)$_{20}$ ± LiN(CF$_3$SO$_2$)$_2$ ± NNP[f]-N(CF$_3$SO$_2$)$_2$	2.8 × 10^{-4}	20	6.20	(376)
P(EO)$_{20}$ ± LiN(CF$_3$SO$_2$)$_2$ ± BMI[g]-N(CF$_3$SO$_2$)$_2$	3.2 × 10^{-4}	25	5.60	(377)

[a]N,N,N-trimethyl-N-hexylammonium
[b]N-butyl-N-ethylpiperidinium
[c]1-methyl-3-propylpyrrolidinium bistrifluoromethanesulfonylimide
[d]2,3-dimethyl-1-octylimidazolium
[e]1-methyl-3-ethylimidazolium
[f]N-methyl-N-propylpyrrolidinium
[g]1-butyl-3-methylimidazolium

4.2.1 Solid polymer electrolytes

Fenton et al. initially discovered that alkali salts could be easily dissolved in poly (ethylene oxide) (PEO) and provide excellent ionic conductivity at temperatures over the crystalline melting point of PEO (391). Solid polymer electrolytes are of two types: homogeneous and heterogeneous. The homogenous electrolytes are based on polymeric solid solution of ions in their pure form, whereas the heterogeneous electrolytes consist of polymers of different phases or structures with multifunctional abilities of ion transport. In general, the structure of polymers is complex in nature, and thus majority of the solid polymer electrolytes are heterogeneous in nature (392). The primary property of any electrolyte system is ionic conductivity, and the range of 10^{-3} S cm^{-1} at room temperature is more preferable for practical applications. The usual low ionic conductivities of polymers (of the order of 10^{-8}–10^{-4} S cm^{-1})

are a challenging factor and the conductivities need to be enhanced (393). The semicrystalline nature of PEO enables formation of different complexes with lithium salts due to the coupling of Li^+ and oxygen moieties of the PEO chains. Albeit these coupling reactions enhance the dissolution of lithium salts, the anchoring effect restricts the effective transportation of ions, leading to the decrease in ionic conductivities.

The amorphous phase of the PEO with activated chain segments showed augmented ionic conductivities, and thus increasing the percentage of amorphous phase is a good solution for improving the conductivities (394). The incorporation of nanoparticles and plasticizers with PEO and blending with other copolymers could improve the ionic conductivities (395, 396). Copolymers with conductive block units are useful for enhancing the ionic conductivities and the formation of self-assembled structures augments the mechanical properties (397). Martin et al. utilized polyvinylpyrrolidone (PVP) as a solid electrolyte for supercapacitor application (398). PVP is an amorphous polymer with moderate ionic conductivity, and it exhibits good stability and high transition temperature (T_g, 170 °C). The high Tg value is due to the presence of rigid pyrrolidone groups, but addition of water can lower the temperature by serving as a plasticizer (T_g, 40 °C). PVP shows Lewis base character due to the presence of tertiary amide carbonyl groups which enable to form complexes with inorganic salts. When PVP is doped with lithium perchlorate, the ionic conductivity increases with respect to temperature and reaches a value of 3.34×10^{-3} S cm^{-1} at 60 °C.

Xu et al. developed a composite solid polymer electrolyte by incorporating $Li_{10}GeP_2S_{12}$ into a PEO matrix for fabricating solid-state lithium batteries (399). PEO has the ability to form solid electrolyte membranes owing to its ability to solvate a range of salts through interaction via oxygen moieties with cations. But its ionic conductivity needs to be improved for productive solid-state battery application. Sulfide electrolytes possess high lithium ion conductivity (10^{-2} S cm^{-1}) and large potential window (>10 V vs. Li^+/Li) (400). Incorporating $Li_{10}GeP_2S_{12}$ in PEO matrix greatly improved the ionic conductivity of solid polymer electrolyte (1.21×10^{-3} S cm^{-1} at 80 °C) and provided a stable wide electrochemical window of 0–5.7 V. When assembled with LiFePO$_4$/Li battery the electrolyte showed good energy storage performance with high capacity retention of 92.5% even after 50 cycles at 60 °C.

4.3 Gel electrolytes

Gel electrolytes combine the advantages of both solid and liquid components and the new state is in between that of individual components. Gel electrolytes possess higher ionic conductivities than that of solid polymer electrolytes and high degrees of interfacial accessibility and low interfacial resistance. Also, gel electrolytes show good mechanical flexibility and enhanced safety which makes them suitable for fabricating wearable energy storage devices. Polar polymers having significant liquid uptake behavior have been utilized as a solid component in gel electrolytes. Polymers like poly(vinylidene fluoride) (PVDF) (401), PEO (402), poly (vinylidene fluoride-hexafluoropropylene) (403), polyacrylonitrile (404), and poly (methyl methacrylate) (PMMA) (405) have showed their liquid affinities and multifunctional properties. In order to achieve a proper balance between the mechanical properties and the ionic conductivities, the gel needs to be structured with the polymer materials that can absorb a large quantity of liquid and maintain the mechanical strength. In addition, the cross-linking of polymers can keep the structural integrity while maintaining the swelling or wettability with an excellent affinity to liquid electrolyte (406). The introduction of nanoparticles could increase the mechanical strength of gel electrolyte and facilitate multiple pathways for rapid ion movements (407). The partial exposure of gel electrolyte to the electrode surface often increases the interfacial resistance and decreases the device performances. In order to achieve complete accessibility of active materials on the porous nanostructured materials, the most facile strategy is to pre-wet the electrodes with the liquid electrolytes. This process maximizes the wetting of electrode surface and provides robust interfacial bonding with gel electrolytes.

For flexible supercapacitors, generally gel electrolytes are prepared by mixing electrically conducting species such as acids, bases, and salts with a water-soluble polymer like polyvinyl alcohol (PVA). Phosphoric and sulfuric acids and potassium hydroxides are frequently used as compounds having conducting ions (408). The supercapacitor performance purely depends on the nature of electrolyte like size and concentration of ions, and solvent concentration. The type of electrolyte and the interfacial interaction with the electrode greatly influence the power density, energy density, specific capacitance, and cycle life of a supercapacitor. Gel electrolytes can be prepared by using both aqueous and non-aqueous electrolytes. Aqueous electrolytes possess high ionic conductivity and low operating voltage due to the electrolysis of water beyond 1.3 V. PVA-based aqueous gel electrolytes are extensively utilized for supercapacitor application (409). Blends of PVA and other copolymers like PVP (410), PEO (411), and PEG (412) are also studied for improving the performances of solid-state supercapacitors.

The non-aqueous electrolytes are majorly divided into three types such as organic, ionic liquid, and mixtures of ionic liquid and organic solvent. Organic solvents such as acetonitrile and propylene carbonate are frequently used to dissolve organic salts, and such electrolytes suffer from poor ionic conductivity but offer high electrochemical potential range of 2.5 to 2.8 V. Anderson et al. utilized a gel electrolyte consisting of PEO and sodium bis(trifluoromethanesulfonyl)imide in a mixture of organic solvents; ethylene carbonate, propylene carbonate, and dimethyl carbonate. The gel-assembled activated carbon supercapacitor showed a stable operating voltage of 2.5 V with high energy density and specific capacitance (413). The poor ionic conductivity can be overcome by using ionic liquid as an electrolyte material. Ionic liquids are molten salts at room temperature and maintain their liquid state. Ionic liquids are non-toxic, environmental-friendly, and provide high electrochemical and thermal stability, and, more importantly, a wide operating voltage of over 3 V. Kim et al. prepared a flexible paper-based supercapacitor using CNTs and ionic-liquid-based gel electrolytes. The electrolyte consists of fumed silica nanopowders and 1-ethyl-3-methylimidazolium bis(trifluoromethylsulfonyl)imide as an ionic liquid. The supercapacitor displayed high power and energy densities at an operating voltage window of 3 V, along with excellent stability and flexibility (414). Similarly gel electrolytes with lithium-based salts such as $LiPF_6$, $LiN(FSO_2)_2$, $LiClO_4$, and $LiPF_3(CF_2CF_3)_3$ are well studied for solid-state lithium batteries (415–418).

4.4 Adhesive composite electrolytes

Achieving an initial interfacial contact between the electrode and electrolyte along with maximum surface wetting can provide appreciable energy storage performances. But it is challenging to maintain the energy storage capacity when the devices are experiencing external stress. The mechanical deformation of electrodes under strain causes delamination or deterioration of electrolyte from the electrode surface. Also, the change in active area/volume of the electrode leads to irregular charge–discharge performances. This issue must be especially critical for wearable supercapacitors and batteries as they come across irregular mechanical deformations during the exercise activities. This is a critical bottleneck for both epidermal electronics and textile electronics where there are high probabilities of non-uniform body motion even during day-to-day activities. The use of gel electrolytes with adhesive behavior can address these issues by facilitating robust anchoring and imperishable ionic pathways.

Zhong et al. developed a gum-like electrolyte to obtain a stable and effective electrode-electrolyte interface under deformation (419). The hybrid electrolyte consists of multi-network structures with a double percolation network. The network of liquid electrolytes is reinforced by a packing layer of solid particles with maximum loading. PEO having high molecular weight with $LiClO_4$ lithium salt was used as polymer electrolyte. The core shell particles with polymer electrolyte as shell was prepared, which offers a strong adhesion of liquid to the particle surface. The hybrid electrolyte became adhesive with more than 40% of the liquid electrolyte, and thus this method is useful for constructing gum-like sticky electrolytes with excellent ionic conductivities. Zhou et al. developed an epoxy-based adhesive polymer electrolyte for supercapacitor application (420). The electrolyte consists of PVDF, lithium triflate, and epoxy, and it showed exceptional ionic conductivity of the order of 10^{-2} S cm^{-1}. The electrolyte was integrated with the carbon-based structural supercapacitor, and it showed good electrochemical performance and interfacial adhesion strength.

4.5 Stretchable electrolytes

The burgeoning area of stretchable electronics necessitates stretchable energy storage device as a power source. The wearable epidermal sensors used for on-body monitoring of healthcare biomarkers are stretchable in nature (337). The stretchable sensors seamlessly merge with the mechanical properties of the skin or other wearable substrates, and their elastic nature mitigates the strain developed during day-to-day activities (421). In order to commercialize the stretchable electronic devices, development of stretchable energy storage devices is much imperative. As discussed in the previous chapters, stretchability can be achieved by two ways: as intrinsically and as structure-dependent. A stretchable energy storage device requires each component to be stretchable, including electrodes, electrolytes, and separator. Stretchable electrolytes offer advanced-level performance, and they withhold high degrees of mechanical agitations to a large extent without any considerable loss in the electrochemical activity. Fabrication of these electrolytes is challenging as their ionic conductivities should not be compensated while maintaining the stretchability. Functional polymers are the integral constituent of any stretchable electrolyte and their elastomeric properties can be enhanced by the addition of nanofillers, plasticizers, cross-linkers, and ionic liquids.

Wallace et al. developed an intrinsically stretchable supercapacitor with a highly stretchable gel electrolyte based on PVA-H_3PO_4 (422). The studies showed that the PVA with a high molecular weight (124,000–186,000 g mol^{-1}) is only providing appreciable stretchability. The film prepared with medium molecular weight PVA (85,000–146,000 g mol^{-1}) possessed extremely sticky character and very poor elasticity during stretch–release cycles. The composite gel of PVA and H_3PO_4 is formed at a weight ratio of 1:1. H_3PO_4 functions as a plasticizer in the PVA matrix, and the free volume of PVA is increased with the addition of H_3PO_4. Also, the low molecular weight PVA shows poor mechanical strength, and, as the molecular weight increases, the overlapping of lengthy polymer chains causes enmeshing each other and forms a stretchable film. When coupled with a stretchable supercapacitor having buckled polypyrrole electrodes, the supercapacitor maintains its performance under 30% strain for 1,000 stretching cycles.

PMMA and its co-polymers show good mechanical strength and good electrochemical properties. Ramaprabhu et al. incorporated ionic liquid with PMMA and used as a stretchable electrolyte for hydrogen-exfoliated graphene-based supercapacitor (380). The addition of [BMIM][TFSI] ionic liquid enhances the ionic conductivity to a larger extent (0.78 mS cm^{-1} at 28 °C), and its plasticizing effect provided fourfold stretchability. The composite electrolyte minimizes the electrode-electrolyte interfacial resistance and realizes very low equivalent series resistance (16 Ω). Polyacrylic acid– and polyacrylamide (PAM)-based hydrogels are quasi-solid-state electrolytes where the extent of cross-linking by hydrogen bonding interaction provides stretchable properties. Albeit these electrolytes offer excellent ionic conductivities, their high water uptake characteristics often ruin their mechanical robustness. This could adversely affect their elastic behavior and result in poor mechanical durability. Zhi et al. developed a hydrogel from agar and PAM that could sustain low residue strain of 5% even after exposed to 500% strain for 30 times (49). The gel consists of double networks from hydrogen bonds involved in agar gel and hydrophobic interaction associated with PAM gel. The elastic gel endows advanced reversible stretchability up to 500%, and it could restore the original length almost 100%. When assembled with soft polypyrrole electrode, the supercapacitor withstands large deformations, and the adhesive nature of the gel avoids its from peeling off from the electrode surface during stretching. The performances of the solid-state supercapacitor are recovered from 100% strain and maintained even after 1,000 stretches.

Graphene oxide (GO) is a promising filler material to facilitate reversible stretchability for nanocomposite hydrogels (423). The two-dimensional GO sheets hold a large surface area, and their basal planes and edges consist of good abundance of polar functional groups such as hydroxyl, carboxylic, and epoxide groups. The presence of functional groups enables facile dispersive interactions with various hydrophilic polymers like dipole–dipole interaction, hydrogen bonding ionic interaction, and many more. Moreover, GO provides an excellent sliding region with frictional interaction that assists reversible movement of polymers during stretch–release cycles. Also, the superior mechanical strength and electron transfer kinetics of GO facilitate synthesizing stretchable nanocomposite hydrogels with good mechanical properties. Introducing chemical and physical anchoring sites on GO offers a synergistic platform for both polymeric interaction and cross-linking, which can greatly improve the mechanical resiliencies. Azam et al. functionalized GO sheets with acrylic groups, and the as-synthesized cross-linker demonstrated good swelling behavior and mechanical properties with polyacrylic acid–based polymeric gels (424).

In lithium-based flexible batteries, it is imperative to use flexible and stretchable gel electrolytes with good mechanical strength, non-flammability, and electrochemical stability. Ionic liquids–based linear polymers suffer from poor mechanical properties due to their limited chain entanglements within the polymer which are separated by small molecules. The mechanical strength of the stretchable gel polymers can be enhanced by high degrees of cross-linking. The cross-linking strategy facilitates interconnecting of polymer chains with the cross-linkers by forming covalent bonds and establishes three-dimensional enmeshed arrangements. Li et al. developed a stretchable gel polymer electrolyte based on cross-linked PVDF-HFP and an ionic liquid, 1-ethyl-3- methylimidazolium bis(trifluoromethylsulfonyl)imide (425). The method is devoid of any molecular initiator, and the polymer chains were cross-linked with triallyl isocyanurate as a cross-linker by electron-beam irradiation technique. Such cross-linked films showed a high level of ionic liquid absorption and thus high ion conductivity. The gel electrolyte possesses high tensile strength (10.6 MPa) and good elasticity with strain recovery limit of 100%.

Majority of stretchable energy storage devices utilize the gel polymer electrolyte as both dielectric separator and ion conductor. But the problems associated with internal shorting cause the entire device failure and major safety issues. The twisting or stretching deformations lead to the delamination or displacement of the gel electrolyte from the electrode surface. Such issues can be addressed by incorporating a physical separator which holds the similar mechanical properties to that of the other components of the energy storage devices. Liu et al. reported a stretchable PU/PVDF separator with adhesive properties via the electrospinning method, and used for stretchable battery application (116). The separator is highly porous, chemically stable, stretchable, and allows rapid Li-ion mobility. These stretchable separators are very useful for realizing all-solid-state reliable and durable energy storage devices that could work well in any complex environment.

4.6 Self-healing gel electrolytes

The wearable energy storage devices with all the components stretchable could alleviate the enormous strain experienced during intense exercise activities and body movements. However, when the device is subjected to abnormal strain beyond the elastic limit, it leads to microscopic-level damages and cracks, and results in poor electrical conductivity and device performances (291). Human skin and some of other biological systems possess the ability of self-repairing of wounds through a sequence of processes. Introducing such self-healing behavior to the device components of stretchable energy storage devices offers long life and stable performances under harsh conditions. Incorporating self-healing ability to the stretchable electrolytes is relatively simple as the presence of ions and liquids in the gel can induce autonomous healing mechanisms (426). The self-healing process quickly restores the original device activity under stretching deformations due to the regeneration of chemical or physical bonds by itself. In addition, the robust performances of the self-healing gel compensates the microscopic level cracks on the electrode surface, and often exhibits enhanced performances due to the increase in active area. The exposure of more active materials through the micro-damages enhances the interfacial area between the electrode and electrolyte, and thus boosts up the energy storage performances. Considerable research interests have been attracted for synthesizing smart supramolecular functional gels with self-healing ability, and its application as gel electrolyte in stretchable supercapacitors and batteries (427–429).

Polyelectrolytes possess satisfactory ionic conductivities in the range of 10^{-4} to 10^{-3} S cm^{-1}, good mechanical properties, and chemical compatibility to interface with the flexible electrodes. Polyelectrolytes and reversible cross-linking agents are largely utilized for synthesizing self-healing ionic gels to function as solid electrolytes in energy storage devices (430). The favorable features required for self-healing electrolyte gel are low glass transition temperature, high molecular weight, low glass transition temperature, high degradation temperature, and rapid segmental movement along the polymer sequences (109, 431). Upon damage, the recovery of microscopic-level internal properties occurs via the reorientation of molecular geometry and structure, which re-generates the ionic conductivities and viscoelastic properties (339).

The self-healing efficiency is qualitatively tracked by many methods like slicing the gel into multiple parts, making a hole, or scraping the surface, and then allowing it to unite together. The quantitative measurements of the self-healing process can be precisely done by the stress-strain analysis, and the tensile strength value indicates the status of self-healing process. But compressive stress-strain study is generally employed for the soft gel evaluation. The self-healing efficiency (ε_h) is determined from the ratio between the elongation distance of cured (λ_b) and the pristine gel ($\lambda_{b,0}$) as represented in equation (4.1) (432):

$$\varepsilon_h = \lambda_b / \lambda_{b,0} \tag{4.1}$$

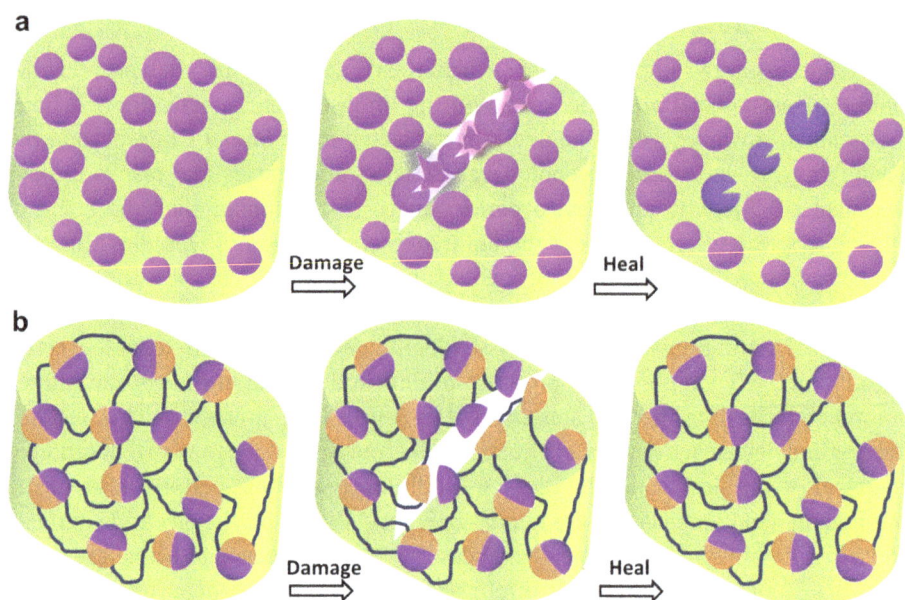

Figure 4.1: Schematic illustration of the self-healing process by (a) the expulsion of healing agents, and (b) the restoration of reversible chemical bonds via supramolecular interactions.

Depending on the accompanying mechanism the self-healing substances are classified into extrinsic and intrinsic materials. Figure 4.1 typically represents the underlying processes involved in both the materials. In extrinsic materials, the self-healing occurs due to the expulsion of healing agent which is preloaded with the bulk of gel electrolyte (Figure 4.1a). Microcapsules are usually utilized to encapsulate organic solvents as the healing agents, and when the gel electrolyte experiences a high level of strain, the capsules break and the solvent facilitates repolymerization at the damaged part and thus repairs the damages (433, 434). The critical shortcoming related to the extrinsic materials is its incapability to perform multiple healing at the same location. Upon releasing of healing agents, the successive repairing is prevented due to the absence of active capsules (435). These bottlenecks can be tackled by introducing intrinsic covalent or non-covalent interactions within the polymeric system that could facilitate consecutive healing even at the same location (Figure 4.1b).

The reversible formation of chemical bonds provides dynamic regeneration of the cracks and restores the ionic conductivity almost instantaneously (436). Depending on the nature of interaction in the self-healing gel electrolyte, the dynamic process can be divided into physical and chemical routes (Figure 4.2). Various non-covalent interactions are involved in the physical healing such as hydrogen bonding (437), π–π stacking (438), metal co-ordination (439), hydrophobic interaction (440), and host–guest bonding (441). The chemical healing depends on the formation of reversible chemical bonds like acylhydrazone bonds (442), imine bonds (443), boronate ester bonds (436), or disulfide bonds (444). The following sections elaborate both of these physical and chemical methods for realizing self-healing gel electrolytes that

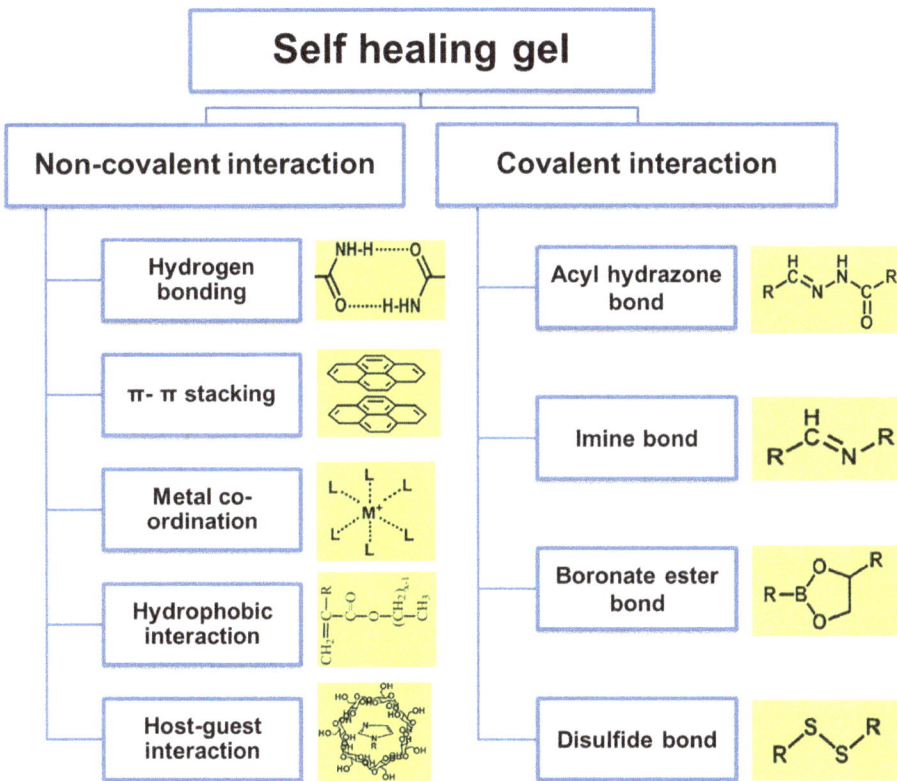

Figure 4.2: Demonstration of various bonding procedures for synthesizing self-healing gel electrolytes.

sustain their performances under abnormal strain. Table 4.2 summarizes various self-healing gels and their properties (444–454).

Table 4.2: Examples of different self-healing gels and their properties.

Polymer	Self-healing mechanism	Solvent	Self-healing efficiency	Self-recovery efficiency	Reference
Cytosine/guanosine-modified hyaluronic acid	Hydrogen bonding	H_2O	–	–	(445)
Carboxymethylcellulose/ Li_2SO_4	Hydrogen bonding	H_2O	92%	–	(446)
PNIPAM-co-PEO derivatives-co -DOPA[a]	Hydrogen bonding, π–π stacking	H_2O	–	100%	(447)
Catechol grafted chitosan/ Fe^{3+}	Metal-coordination	H_2O	–	100%	(448)
PHEA-API[b]/Ni^{2+}	Metal-coordination	H_2O	–	100%	(449)
Gelatin	Hydrophobic interactions	Ionic liquid	96%	–	(450)
Glycyrrhetinic acid/β-CD	Host–guest interactions	H_2O	100%	–	(451)
Tetra-PEG derivatives	Acylhydrazone bonds	H_2O	–	–	(452)
Polydopamine derivatives	Imine bond, Michael addition	H_2O	–	>95%	(453)
Boronic acid-modified alginate	Boronate ester bonds	H_2O	100%	–	(454)
Poly(PEGMA)[c]/BMOD[d]; poly (HEAA[e]-co-META[f])/BMOD	Disulfide bonds	H_2O	–	–	(444)

[a]poly[(N-isopropylacrylamide)-co-(N-3,4-dihydroxyphenethyl acrylamide)]-b-poly(ethylene oxide)-b-poly[(N-isopropylacrylamide)-co-(N-3,4-dihydroxyphenethyl acrylamide)]
[b]poly(2-hydroxyethyl-co-1-(3-imidazolyl) propyl aspartamide);
[c](poly(ethylene glycol)methyl ether methacrylate
[d]bis(2-methacryloyl)oxyethyl disulfide
[e]N-hydroxyethyl acrylamide
[f]2-(methacryloyloxy)ethyl trimethylammonium chloride

4.6.1 Self-healing gels based on non-covalent interactions

The autonomous recovery of mechanical and rheological properties of materials after being wounded has attracted pervasive attention, and tremendous efforts have been paid to design and synthesis of efficient self-healing materials. Despite the advances only very few types of self-healing materials are successfully used for energy storage applications. This is due to the challenging factors such as thermal stability, mechanical resiliency, electrical and ionic conductivities, electrochemical activity, and processability.

4.6.1.1 Hydrogen bonding

Hydrogen bonding is a versatile supramolecular interaction which forms intrinsically self-healing gels by cross-linking of small molecules. Such interactions provide dynamically self-healing electrolyte with good mechanical properties and superior viscoelastic or thixotropic behavior, and water uptake properties. Dual physical cross-linking is appropriate for synthesizing hybrid hydrogels with double-networks, which can withstand high levels of strain. Combination of ionic interactions and hydrogen bonding is a facile approach for realizing cross-linked network of homogenous gels that exhibit an energy dissipation mechanism. When the cross-linkages are damaged, instantaneous self-healing happens by rapid regeneration of ionic and hydrogen bonds.

Li et al. developed a flexible supercapacitor having physically cross-linkable polyelectrolyte based self-healing gels (384). The copolymerization involves hydrophilic and hydrophobic monomers like acrylic acid and stearyl methacrylate, respectively. The hydrogel exhibits high ionic conductivities (>30 mS cm^{-1}) and good self-healing and stretching efficiencies. The presence of movable protons provides high capacitance to the PANI nanowire supercapacitor with an energy density of 19.33 Wh kg^{-1}. The polyelectrolyte gel possesses high extensibility ($>2,400\%$) and multiple self-healing ability with a capacitance retention of 86%. Polyacrylamide chitosan–based supramolecular gel also showed substantial stretchable and self-healing properties (455). The hydrogel consists of cross-linked networks of inter-chain hydrogen bonds between short chitosan chain and polyacrylamide through free-radical polymerization. The gel withstands successive compressions – relaxation strain cycles, and can be stretched to 30 times of its original length. Ureidopyrimidinone units have recently emerged as a hydrogen-bonded gel electrolyte formed via quadruple hydrogen bonding. Xue et al. developed a stretchable battery based on cross-linked polyelectrolyte having ureidopyrimidinone and brush-like poly(ethylene glycol) chains. The system can be stretched to 20 times of its length, and after cut damage, the autonomous self-healing happens within 2 h at ambient conditions (456).

4.6.1.2 π–π stacking

Polymers cross-linked by weak dynamic bonds exhibit soft and viscoelastic properties, and the incorporation of reversible bond formation abilities provides additional self-healing properties. The π-electron rich polymeric gels endow self-healing activity via π–π stacked supramolecular interactions. Pyrene and naphthalene diimides possess π–π stacking interactions and extensively utilized for formulating autonomous healable materials. In general, π-electron rich species forms stable supramolecular complexes with π-electron deficient materials. Perylene forms stable complexes with macrocyclic aromatic ether imide sulfones (457) Hayes et al. reported a self-healing system using π-electron rich bisperylene-terminated polyether with a π-electron-deficient chain-folding polydiimide. The polymer system provides a tensile modulus of 10 MPa along with 100% recovery over three break/heal cycles.

Pt-Pt interaction is a promising metallophilic interaction existing in platinum(II) atoms and the interaction influences structure and photo physical properties of the Pt complexes. The strength of Pt(II)-Pt(II) interaction is around 10 kcal mol^{-1}, and the bonding can be strengthened by introducing π–π stacking interaction. Platinum(II) complexes show reversible association–disassociation process with appreciable dynamic behavior. Hence, when polymers are cross-linked by these intermolecular interactions between platinum(II) complexes, the as-formed gel electrolytes exhibit good elasticity and intrinsic self-healing abilities (458).

4.6.1.3 Metal co-ordination

The inherent reversibility of properly tuned co-ordination cross-linkages facilitates the formation of self-repairing hydrogel materials, which remain intact even after getting ruptured or sliced. Hydrogels based on metal–ligand coordination are mostly achieved by functionalizing polymer backbones with specific chelating ligands, and when treating with metal ions they form cross-linked gels. The metal center donates an electron pair to a ligand and establishes a coordination bond. The binding strength depends on the nature of the metal ions and the coordination number. These dynamic hydrogels also possess a shear thinning behavior, which means the viscosity decreases under shear stress. By tuning the metal–ligand co-ordination chemistry, reversible polymeric networks with good mechanical strength can be constructed.

Various chelating ligands are shown to have reversible co-ordination like catechol, bisphosphonate, thiolate, histidine, carboxylate, pyridines, and iminodiacetate conjugated onto polymeric platform, and many of these chelated dynamic networks are used as self-healing electrolyte in flexible energy storage devices (459). Zhai et al. developed a supramolecular hydrogel through non-covalent cross-linking of adenosine 5-monophosphate with Mn^{2+}(460). The hydrogel is cross-linked with PVA and the hybrid polymer networks hold good mechanical properties. Owing to the self-healing ability, mechanical strength, and stability, the gel electrolyte is utilized to fabricate stretchable Zn-MnO_2 battery. The multifunctional gel eliminates the dissolution of electrode materials, and provides fast ion conductivity and excellent electrochemical stability. Such metal–organic supramolecular hydrogels offer a promising opportunity for wearable energy storage application.

4.6.1.4 Hydrophobic interaction

The dynamic micelle formation mechanism is appropriate for synthesizing self-healing gel electrolytes (461). The introduction of hydrophobic functional groups over long alkyl side chains of hydrophilic polymer establishes an energy dissipation mechanism and results in soft gel materials with significant rheological properties. Above a particular level of concentration, the hydrophobic groups in the polymers establish intermolecular associations that function as reversible cross-linkage within a 3D polymer matrix. The free radical micellar polymerization method is suitable for preparing such associate polymers having dynamic self-healing, elasticity and water uptake ability. Acrylamide is one of such hydrophilic monomer that can be copolymerized with a water-insoluble hydrophobic monomer via micelle formation. The high concentration of hydrophobe inside the micelle facilitates the hydrophobic monomers to distribute as random blocks over the hydrophilic backbone. But large hydrophobes like dococyl acrylate or stearyl methacrylate are not able to solubilize within the micelles, and thus difficult to establish such dynamic hydrophobic interactions (440).

The addition of surfactant like sodium dodecyl sulfate (SDS) enables to solubilize large hydrophobes in a micellar solution containing an electrolyte, such as NaCl, where the salts facilitate growth of micelles and high degrees of solubility (440). One of such self-healing gel was developed by Okay et al., where the reversible disentanglements of the hydrophobic building blocks establish the physical cross-links, which give rise to a self-healing efficiency of almost 100%.

4.6.1.5 Host–guest interaction

Host–guest interactions are extensively utilized for synthesizing supramolecular gels and supramolecular polymers. Host–guest complexes provide fixed host–guest directionality and geometry, and they can function as multifunctional materials with exceptional mechanical properties, shape memory, self-healing, and expansion–contraction characteristics. When the host–guest self-healing gel is damaged, the complex is damaged into dangling-free host and guest moieties. The individually broken host and guest molecules instantaneously recognize each other and rejoin, which results in autonomous self-healing of the gels.

Huang et al. developed crown ether–based supramolecular polymer gels by coupling a poly(methyl methacrylate) polymer having pendent dibenzo[24]crown-8 groups and two bisammonium cross-linkers with different end-group sizes (462). The rheological studies showed that the host–guest gel possess 100% recovery in less than 10 seconds for several cycles even under a very large strain of 10,000%. Harada et al. showed another example of supramolecular gel having two different kinds of guest moieties: one is an inclusion complex between β-cyclodextrin and adamantane, and the other is between β-cyclodextrin and ferrocene. The selective molecular recognition properties of the gel facilitate multiple functionalities like self-healing, shape memory, and expansion–contraction properties.

4.6.2 Self-healing gels based on covalent interactions

Dynamic covalently constructed hydrogels undergo structural changes in response to an applied stimulus due to the reversible formation and cleavage under specific conditions. Dynamic hydrogels and their associated chemistries are getting increased attention for preparing self-healing hydrogels with appropriate physical and rheological properties. Some of the self-healing gels synthesized via various dynamic covalent chemistries are described in the following sections.

4.6.2.1 Acylhydrazone bonds
Acylhydrazone bond is formed by the condensation reaction of hydrazide with carbonyl group, and the bond is reversible under mild conditions; in the presence of acid catalyst the bond easily regenerates to the starting reagents (463). Polyacylhydrazones are the dynamers which show the capability to exchange their components with external monomers, and the chemical modification occurs almost spontaneously. The acylhydrazone moiety provides the dynamic character via the reversible nature of imino bond and hydrogen bonding spots through the amide group. The reaction of dihyrazides with dicarbonyl compounds yields polyacylhydrazone polymers through a polycondensation reaction (464). Chen et al. developed a cross-linked polymer gel having reversible sol-gel transition and self-healing characteristics. The dynamic polymer gel with a network of acylhydrazone cross-links is prepared by condensation of acylhydrazines at the two ends of PEO with aldehyde groups in tris[(4-formylphenoxy)methyl]ethane. The self-healing behavior enables the polymer gel to rapidly merge together after being sliced into multiple pieces (465).

4.6.2.2 Imine bonds
The imine bond is a dynamic covalent bond which is formed between the aldehyde groups and the amino moieties, resulting in Schiff base formation. Imine bond possesses the bond dissociation energy of 147 kcal mol^{-1}, which attributes rapid degenerative bond exchange without involving any considerable side reactions. The formation of reversible imine bond allows for rapid regeneration of mechanical contacts at the fracture surfaces and facilitates rapid multiple self-healing without the aid of any external stimuli. Wei et al. developed a self-healing chitosan hydrogel

based on imine bonds (466). Chitosan possesses good biocompatibility and the presence of abundant amino groups on its backbone allows for rapid construction of injectable self-healing gels. Aldehyde groups from polysaccharides, di- or multi aldehydes can be used as a cross-linker and their mechanical properties can be controlled by tuning the end or side chains.

4.6.2.3 Boronate ester bonds

The complexation between boronic acids and 1,2- or 1,3-diols forms reversible boronate ester bonds that can be used for preparing cross-linked self-healing hydrogels. The pH and pK_a of boronic acid greatly influence the reversibility and strength of the boronate ester bonds. The boronate ester linkage is favored at pH values greater than the pK_a of the boronic acid; the low pH attributes the formation of free boronic acid and diol. The dynamic equilibrium between boronate esters and boronic acids/diol facilitates bond rearrangements and the self-healing processes (467). It is important to reduce the pK_a of phenylboronic acid–based materials that promote the formation of dynamic boronic esters with good reversibility under physiological conditions. Introducing electron-withdrawing groups to phenylboronic acid can significantly lower the pK_a values. Tan et al. developed boronic ester–based self-repairing bulk hydrogels via B-N intermolecular co-ordination (468). The hydrogel was judiciously synthesized by mixing boronic acid–containing poly(acrylamide-*co*-2-acrylamido glucose) and diol-containing poly(acrylamide-*co*-*N*-acryloyl-3-aminophenylboronic acid-*co*-*N*-(3-dimethylaminopropyl) acrylamide). The incorporation of amine-containing *N*-(3-dimethylaminopropyl) acrylamide increases the stability of the reversible cross-links at physiological pH. The gel possesses good viscoelastic properties (100% strain) and self-healing behavior, and after healing it can be stretched to a length several times of its original length without any breakage.

4.6.2.4 Disulfide bonds

Disulfide chemistry is quite versatile for obtaining self-healing hydrogels by exploiting weaker disulfide bonds. The disulfide groups enable self-repairing at ambient temperature, and these dynamic bonds are highly sensitive to various external stimuli such as pH, redox agents, light, heat, and mechanical stress (469). In addition, the disulfide groups can be easily broken by a reduction reaction forming two thiol groups and re-formed by an oxidation reaction. Jeon et al. reported an attracting strategy for developing multifunctional hydrogels by copolymerizing meso-2,3-dimercaptosuccinic acid and 2,3-dimercapto-1-propanol (470). The presence of network of dynamic poly(disulfide) bonds on the polymeric backbone provides rapid reversible cross-linkages. The hydrogels display various functionalities like fast self-healing (within 5 s), tremendous stretchability (~4,500%), three-dimensional printability, and good electrical conductivity and biocompatibility. These characteristics make it suitable for functioning as gel electrolyte in wearable energy storage devices.

4.7 Conclusion

Flexible and stretchable energy storage devices such as batteries and supercapacitors are required for powering wearable electronics. In order to realize completely stretchable devices, it is important to rely on stretchable and self-healing gel electrolytes. The present chapter outlined the recent advances in smart multifunctional gels for not only functioning as an electrolyte, but also maintaining the electrodes and augmenting the device performances. The various types of electrolytes based on their physical states and the use of solid electrolytes and their additives are discussed. The various synthesis protocols of stretchable and self-healing gels, the mechanical and ion transport properties, and their application as an electrolyte medium in supercapacitors and batteries are detailed. The gel electrolytes enhance the mechanical endurance, fatigue resistance, recyclability, and cycling stability of the overall performance of the next-generation wearable energy storage devices.

5 Hybrid wearable energy storage–harvesting devices

5.1 Introduction

Wearable electronics are becoming indispensable in our daily life in enriching our recreational and professional lives. The emerging trends are focusing on designing these electronics to be smarter with multifunctional features within a miniature device (471). Such portable electronics are extensively used for sensing physiological parameters, healthcare or fitness monitoring, navigation, and wireless communication. These devices are generally integrated on wearable platforms such as directly on skin, textile substrate, bracelet, and eyeglasses (19). The biggest challenge associated with these personalized electronics is integrating them with sustainable power sources (472). Majority of these devices are required to operate continuously, especially for healthcare- or lifestyle-monitoring application. Also, wireless data transmission is an integral part of all smart devices, which requires a large amount of energy. Wearable energy storage devices are appropriate for functioning as a power source; especially LIBs are lightweight and offer high energy density. These energy sources should possess flexibility, stretchability, and self-healing ability so that they can adapt well with any wearable platforms (473).

Although the existing state-of-the-art batteries provide sufficient energy density to power smart wearable devices, the longtime continuous operation often necessitates intermittent recharging of the batteries (474). The common way to enhance the energy density of batteries is to increase the loading of the active materials. But this could eventually increase the size and weight of the energy sources, and that is not appropriate for wearable application. Wearable devices are expected to be thin, miniature, and portable, and should merge well with the wearable substrates without adversely affecting the day-to-day life of the individuals (475). Significant research attempts have been dedicated on developing nanostructured electroactive materials to enhance the energy densities of supercapacitors and batteries (476). Despite these innovations, the existing wearable electronics are hardly meeting the power requirements. Another efficient strategy is to integrate the energy storage devices with energy generation systems. Such amalgamation facilitates efficient charging, and the hybrid system can functions as self-powered platform for wearable electronics (477, 478).

The wearable energy-harvesting component captures energy from the natural sources like solar, heat, or human movements and converts into usable energy that can be stock up in various energy storage platforms (479, 480). Individually, the energy harvesters are not stand-alone because these resources are not available always, and thus the energy production is limited to when and where the resources are available. Hence the energy storage devices can function as energy buffering

https://doi.org/10.1515/9781501521287-005

components for continuously powering the wearable electronics by compensating the intermittent power production from the energy harvesters (481). These self-sufficient power systems serve as a stable power source with a very long operational time, and eliminate the frequent charging requirements (472). Several hybrid systems are successfully developed for exploiting the environmental energies such as photovoltaic devices for solar energy (482–485), piezoelectric nanogenerators (PENGs), and triboelectric nanogenerators (TENGs) for vibrational and mechanical motion (486–490), and thermoelectric and pyroelectric devices for thermal energy (491–495). Biofuel cells convert the abundant chemical energy from molecules to electrical energy with the help of efficient electrocatalysts. Several metabolites available in sweat are useful energy sources, and the biofuel cells can catalyze the energy conversion processes (108, 421, 496–499). As illustrated in Figure 5.1, the energy produced from these epidermal energy harvesters can be stored into the wearable energy storage systems like batteries and supercapacitors. Such amalgamation facilitates to improve the energy densities, and eliminate the energy buffering problems and frequent charging requirements. The following sections of the chapter discuss the various approaches involved in developing such self-sufficient hybrid power sources that could be used for self-powered wearable applications.

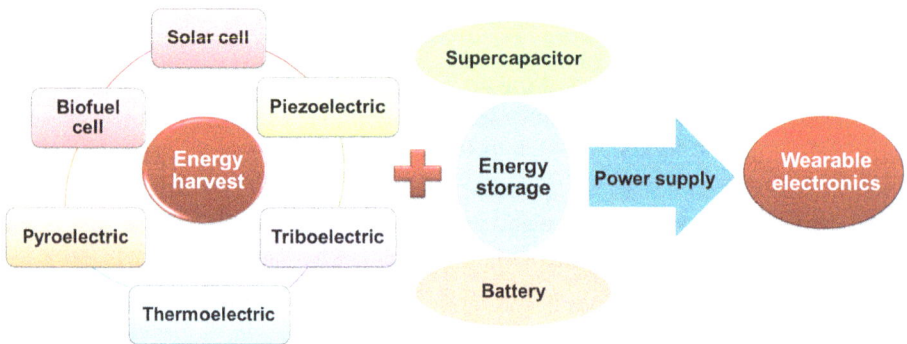

Figure 5.1: Schematically illustrating various energy harvesting and storage devices for self-powering wearable electronics.

5.2 Solar cell battery/supercapacitor

Solar energy is one of the most promising renewable energy sources due to the unlimited source of energy and sustainability, and they are free from any toxic pollution problems (491). The most efficient way to harvest solar energy is to convert it into electrical energy using solar cells. The working mechanism behind the solar cell is the photovoltaic effect which is the generation of voltage and electric current from a material when it is exposed to light. Several processes are involved when light is incident on a solar cell such as the light absorption by a semiconductor material, hole/electron separation, and charge transport to the electrodes. Significant studies have been focused on to enhance the efficiency of solar cells. The conversion efficiency (η) of a solar cell reflects the cell's performance and can be defined as the ratio of output energy from the solar cell to input energy from the sunlight. The efficiency of the solar cell depends on the spectrum and intensity of the sunlight, and the temperature level of the cell. The efficiency of a solar cell is calculated by the fraction of the incident power which is converted into electricity and can be represented as equation (5.1):

$$\eta(\%) = \frac{P_{max}}{P_{in}} = \frac{V_{oc} \times J_{sc} \times FF}{P_{in}} a \times 100 \qquad (5.1)$$

Where V_{oc} represents the open-circuit voltage, J_{sc} is the short-circuit current, and FF is the fill factor. The open-circuit voltage represents the maximum voltage accessible from a solar cell, and this occurs at zero current. The short-circuit current is the current that can be drawn from the solar cell when the voltage across the solar cell is zero.

There are different types of solar cells including dye-sensitized solar cells (DSSC), polymer solar cells, perovskite solar cells, and silicon solar cells. The DSSC consist of five components such as a layer of transparent conductive oxides coated on a mechanical support, a semiconductor film (typically TiO_2), a sensitizer functionalized over the semiconductor surface, an electrolyte having a redox mediator (usually I^-/I_3^-), and a counter electrode for regenerating the redox mediator (500). In DSSC, the charge generation happens at the semiconductor–dye interface, and both the semiconductor and electrolyte facilitate the charge transport. The optimization of the spectral characteristics can be made by modifying the structure of dye alone, but the carrier transport features can be augmented by tuning the composition of the semiconductor and the electrolytes (501).

Polymer solar cells possess distinctive advantages like low specific weight, tunable material characteristics, and low cost. A typical polymer solar cell consists of a transparent conductive layer, usually indium tin oxide as cathode, a metal anode having low work function, and an interfacial layer composed of a conjugated polymer and a fullerene derivative to function as a donor and an acceptor, respectively. The cathode is commonly functionalized with n-type metal oxides (502), PEDOT/PSS (503, 504) or metal carbonates (505). Perovskite was initially used as a sensitizer in DSSCs in place of the molecular dyes. Perovskite solar cells can function even in the absence of an electron injecting layer, and their efficiencies have increased from 3.8% in 2009 to 25.5% in 2020 in single-junction architectures (506).

The energy produced from the solar cells can be directly used to power the wearable electronics, and in the meantime the excess energy can be stored in the integrated electrochemical energy storage devices such as batteries or supercapacitors. The stored energy can be used for powering the personal electronics in all indoor environments and in the scenarios where the sunlight intensity is very low or not available. Similar to energy storage devices, all the wearable energy harvesters should possess superior mechanical properties like flexibility, bendability, foldability, stretchability, and self-healing ability so that they can conformally merge with the soft skin or textile substrates and maintain their performances under mechanical deformations. Elastic conductors are vital components for all stretchable devices and various strategies are available to realize such current collectors. Constructing conductive metal films on elastic polymers or fiber substrate with fracture or buckling configuration is an efficient method (507). Conducting polymers-based 2D buckled stretchable electrodes has been shown to have exceptional stretchability and used for battery application (508). In this approach, polypyrrole was electropolymerized on pre-strained Au-coated elastic substrate and utilized for preparing Mg alloy–based biocompatible battery. The battery endured 2,000 repeated stretching cycles under 30% tensile strain and showed stable electrical conductivity and energy storage properties.

The most important component for all wearable devices is the current collector, its mechanical properties greatly influences the overall performance of the devices. Textile offers a large surface area for all wearable devices, including electronic controller and wireless data receiver/transmitter. The wearable e-textile technologies are exponentially growing in the past decade, and they are able to perform various functionalities such as healthcare monitoring, heat regulation, and luminescent display. Developing flexible conductive yarns with diameters comparable to that of conventional fabric yarns is essential for fabricating textile-based wearable devices. In addition, the common non-conductive pure textile threads can be coated with conductive metals to realize conductive yarns. Coating processes like electroless plating, sputtering, chemical vapor deposition, and dip coating are appropriate for establishing conductive polymer coating. These conductive threads can be tirelessly merged into textiles using weaving, embroidering, breading, or knitting machines (509). Choi et al. exploited a textile substrate and developed a wearable battery (Figure 5.2a), rechargeable by solar energy (510). Nickel-coated textile thread possesses good electrical conductivity and the use of stretchable binder like PU resulting in a current collector that withstands high levels of mechanical deformations, including multiple stretch-release cycles. PU-based porous polymer film is suitable for functioning as a stretchable separator to avoid electrical contact between the electrodes. The frequent separate charging of the fabric based battery is inconvenient and cumbersome. Integrating energy harvesters like flexible solar cells into the textile battery facilitates effortless recharging and avoids the inconvenience of wiring the battery to an external power outlet. In the discharging mode, the battery functions as an energy source, and in the solar-charging mode, the battery is charged (Figure 5.2b). The hybrid system can power nine light-emitting diodes (power consumption of each LED = 0.042 W), simultaneously (Figure 5.2c).

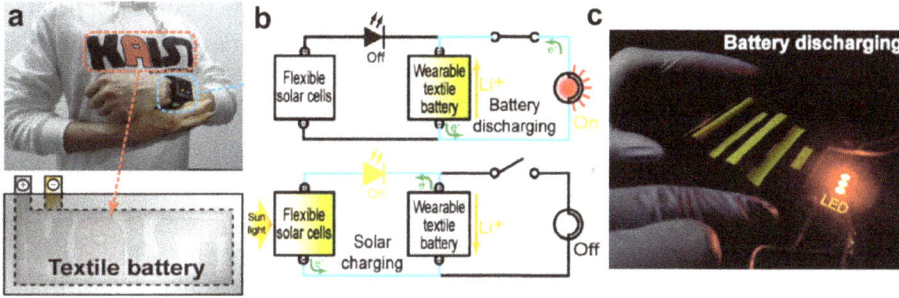

Figure 5.2: (a) A photograph of wearable textile battery integrated in fabric. (b) Equivalent circuit diagram of a solar-powered textile battery in the discharging and solar-charging modes. (c) Photograph showing the lighting up nine LED bulbs by the solar-charged textile battery (reproduced with permission from ref. (510)).

Transparent conductive substrates are needed as current collector for solar cell applications. Indium tin oxide (ITO) is a good choice for transparent conductive electrodes owing to its excellent conductivity, transparency, and high optoelectronic performances. ITO-coated flexible plastic films like PET and PEN are versatile substrates for all wearable photovoltaic applications. Table 5.1 elaborates the vari-

Table 5.1: Various methods for the fabrication of photo-electrodes based on flexible conductive substrates and their cell performances.

Substrates	Active material	Deposition technique	Post-treatment	Efficiency (%)	Reference
ITO-PET	TiO_2	Doctor blading	Mechanical pressing	2.3	(511)
ITO-PET	TiO_2	Hydrothermal crystallization	Autoclaving 100 °C 12 h	2.5	(512)
ITO-PET	TiO_2	Spray deposition or pulsed laser deposition	UV, 248 nm pulse width 20 ns	3.3	(513)
ITO-PET	TiO_2	Electrophoretic deposition	UV, 254 nm	3.8	(514)
ITO-PEN	TiO_2	Doctor blading or screen printing	Heating 110–125 °C	5.8	(515)
ITO-PEN	TiO_2	Doctor blading	Mechanical pressing	8.1	(516)
Ti	TiO paste	Screen printing	Sintering 325–500 °C	7.2	(517)
Stainless steel	TiO_2	Doctor blading	Sintering 600 °C	8.6	(518)

ous methods utilized to prepare flexible photo-electrodes on flexible conductive substrates (511–518). One-dimensional nanostructure such as nanowires, nanorods, and nanotubes are utilized for producing highly efficient solar cells (519). Compared with bulk active materials, 1D nanomaterials possess high specific surface area and exceptional charge transport kinetics. The mechanical properties of fiber-shaped flexible electronic devices are superior to the conventional planar-structures, and the fibers can be woven or knitted into fabric substrates (520). Peng et al. developed a stretchable DSSC based on elastic and electrically conducting fibers (16). The fiber electrodes are prepared by winding aligned MWCNTs on elastic rubber fibers, which hold good electronic properties even under stretching. In order to construct the wire-shaped DSSC, a modified Ti wire was twisted on the elastic MWCNT fiber and then coated with photoactive materials. The fibers can be woven into any stretchable, photovoltaic textile with a desired configuration and the device demonstrated large energy conversion efficiencies (7.13%) under a strain of 30%.

The energy harvesting from the solar cell is limited to the presence of light and illumination intensity. Hence, the availability of solar energy strongly depends on the weather conditions, and the energy-harvesting possibility is restricted to outdoor regions. In order to overcome these limitations of unpredictable and intermittent nature of solar energy, different types of energy harvesters need to be combined. The simultaneous scavenging of energies from different sources provides a reliable platform for realizing self-sufficient wearable electronics. Wang el al. developed a prototype of a textile-based hybrid self-charging system by harvesting solar energy from ambient light and mechanical energy from human motion (521). The energy gathered from the fiber-shaped hybrid system is used to charge an RuO_2-based fiber supercapacitor. The fiber shape of the all-in-one energy harvesting–storage platform facilitates it to be woven into electronic textiles and to fabricate smart clothes.

5.3 Piezoelectric-battery/supercapacitor

Piezoelectric energy harvesters convert mechanical energy from the stress/strain into electrical energy by piezoelectric effect. The sources of mechanical energies can be from body motion, vibrations, air flow, and acoustic noises. The piezoelectric effect is a coupling phenomenon, results from the sequential electromechanical interaction among the mechanical and electrical states in crystalline materials having no inversion symmetry. Upon application of mechanical stress, the crystal structure of the piezoelectric material is deformed and this facilitates the movement of electrical charges. The piezoelectric effect is a reversible process, and the application of electric field to a piezoelectric crystal results in internal generation of mechanical strains. For instance, lead zirconate titanate crystal can generate measurable piezoelectricity when its crystal structure is stressed by 0.1% of the original dimension. But, when an external electric field is applied to the crystal, its dimension changes to about 0.1%. This inverse piezoelectric effect is generally utilized for the production of ultrasonic sound waves.

The polarization charge density is proportional to the applied mechanical stress as per equation (5.2), and the charge density arises from the electric field and potential is represented as equation (5.3).

$$\rho = dX \tag{5.2}$$

$$\Delta E = \frac{\rho}{\varepsilon} \tag{5.3}$$

where ρ indicates the polarization charge density, d is the piezoelectric coefficient, X denotes the applied stress, ΔE is the divergence of the electric field and ε represents the permittivity. Non-ferroelectric crystals like quartz naturally exhibit stable piezoelectric effect but the effect is not very strong. The external mechanical stress causes distortion of the crystal lattice that induces the formation of electrical dipoles. In ferroelectric materials, the piezoelectricity can be generated through poling. The ceramic piezoelectric transducer is usually produced by pressing microstructures of ferroelectric materials. The powder needs to be heated above Curie temperature and once it cools down, the perovskite structures undergo phase transformation. The conversion from the paraelectric to the ferroelectric state results in randomly oriented ferroelectric domains. This random orientation causes zero net polarization and piezoelectric coefficients. Spontaneous polarization occurs more easily in perovskite type crystal structures, where the polarization appears without the aid of an external electric field. In the absence of spontaneous polarization, the material needs to be polarized via poling.

Wang et al. utilized the piezoelectric effect by generating electricity from the piezoelectric material in combination with the electrodes, and the resulting piezoelectric nanogenerator (PENG) was used as a source of electrical energy (522). PENGs have met significant attraction from the researchers, and various nanostructures based prototypes such as vertically grown and laterally aligned are developed (523, 524). Semiconducting (525, 526) and insulating materials (527) have also shown promise for piezoelectric energy-harvesting application. PENGs can be used to convert different forms of mechanical energies such as pressing (528), vibration (529), rain drop (530), sound (531), bending (532), stretching (533), muscle movements (534), heartbeat (535), respiration (536), wind (537), and fluid flow (538) into electrical energy.

The biointegrated wearable devices such as pacemakers, heart rate monitors, neural stimulator, and cardioverter defibrillators provide continuous diagnosis and therapy, and require wearable power sources for their incessant operation. The battery life of implantable devices is much crucial as it necessitates surgical procedures to replace the batteries, which may lead to potential health risks and even mortality. Integrating wearable batteries with appropriate energy-harvesting sources is a good solution to extend their lifetimes significantly. Regular human body motions like cardiac and lung movements serve as inexhaustible energy sources throughout the lifetime of a patient. An ideal piezoelectric device could harvest energy during the macroscale displacement of natural motions of an organ, but it should not cause any constraints to those motions or functions. Flexible and stretchable piezoelectric devices can be seamlessly merged with the human body as a secondary skin and can harvest energy from the natural relaxation and contractile motions of the heart, lung, and diaphragm. Rogers et al. demonstrated such a hybrid energy-harvesting platform in several different animal models and co-integrated with rectifiers and microbatteries. The bioelectronic system provides sufficient power for operating pacemakers over 20 million cycles of bending/unbending in a hydrogel environment (539).

Human body is an abundant source of energy, including both chemical and physical energy. The kinetic energy is accessible from various parts of a human body, and piezoelectric shoe-based energy harvesting is appropriate for wearable devices (540). The walking activity provides three harvestable sources of energy such as acceleration pulse from heal strike, leg swing, and the compressive force due to the weight of an individual. Majority of the shoe-based harvesters developed to date can only harness energy from a single excitation. Some of the PENGs utilize multidirectional vibrations, but the excitations from various directions possess low frequencies. Hence, the frequency up-conversion technique, hybrid energy-harvesting design, and the method to superpose different excitations are needed to exploit the walking energy effectively. Wang et al. developed a self-charging power system by combining energy-harvesting PENG and energy-storing LIBs. Instead of integrating two different systems, both the mechanisms can be directly hybridized into a single platform where the converted mechanical energy can be simultaneously stored. The use of PVDF separator between the electrodes of a LIB facilitates utilizing the piezoelectric potential from a PVDF film for efficient Li ion migration from the cathode to the anode and the associated charging reactions. Such a flexible self-charging platform withstands high levels of strain developed during bending and twisting deformations (541).

5.4 Triboelectric battery/supercapacitor

Triboelectric nanogenerators (TENGs) are working based on two phenomena such as contact electrification and electrostatic induction. The contact electrification is a process of movement of charges from one side to another while two materials come into contact (542, 543). The electrification is considered as a negative effect due to the generation of high voltages but can be exploited as a reliable energy source with the help of TENGs. When two different materials are sliding or contacting through external mechanical energy, the electrostatic charges are induced on both the materials. When the mechanical force separates the material, electric potential is generated and the charges tend to move from one side to another. The charge transfer in this triboelectrification is not exactly reversible, but the excess and deficit of charges are created on the opposite sides, respectively. Therefore, an alternating current flow occurs based on the charge polarity. The generated potential V and the current I across an external load are determined using equations (5.4) and (5.5), respectively:

$$V = -\frac{\rho d}{\varepsilon_o} \tag{5.4}$$

$$I = C\frac{\partial V}{\partial t} + V\frac{\partial C}{\partial t} \tag{5.5}$$

where ρ represents the triboelectric charge density, ε_o is the vacuum permittivity, d indicates the interlayer distance in a particular state, C is the capacitance of the system, and V denotes the voltage across the electrodes.

Several mechanisms have been postulated for the underlying charge transfer processes; an electron transfer mechanism is the most accepted that is based on the surface trap states at metal–insulator and insulator–insulator contact regions. The charge transfer magnitude for the interfacial pairs like metal–metal, semiconductor–semiconductor, and semiconductor–metal is strongly related to the corresponding work function differences (544). Ion transfer mechanisms also happens in some of the contact electrification processes where mobile ions are transferred from one material to another during their physical contact. The asymmetric separation of these charges leads to the triboelectric process. For instance, in ionic polymers, where one ion is anchored with a polymer matrix and another ion having a different polarity is mobile, the charge separation is critically governed by the movement of mobile ions (545). Some attempts have been focused on understanding the correlation between the charge transfer and material transfer in polymer materials. The triboelectrification in polymer materials via material transfer is often accompanied by bond cleavages (546).

TENGs are promising for scavenging minute mechanical energy and converting into large-scale electrical energy owing to their high effectiveness, universal availability, low cost, and environmental friendliness. TENGs can be used to harvest energies from human body motion (547), wind (548), rain drops (549), water waves (550), and many more. For self-powered wearable devices, it is necessary to design the TENGs as mechanically flexible, soft, light, and stretchable. Wearable textile-based TENGs should possess additional features like being washable and breathable, and thus they can be directly used as a cloth. Wang et al. developed a textile-based self-charging power unit by integrating a TENG unit with a flexible LIB belt. Electroless plating is a good method to convert the insulating polyester fabrics into a conductive substrate which can function as a current collector. The TENG cloth can harvest energy from regular human motions when worn at different regions of the body. A Ni-coated polyester textile-based LIB was developed with LiFePO$_4$ and Li$_4$Ti$_5$O$_{12}$ as active materials, and they can be folded at 180° for 30 times. This hybrid platform can power a heartbeat sensor capable of wireless communication that shows its viability for smart wearable electronics (551).

The wind-induced vibrations can be utilized to harvest energy from the natural environments. Wind energy is abundant and available from natural as well as synthetic sources. Integrating LIBs with these TENGs facilitates periodic charging, and the coupled system can be used for powering wearable devices. Yang et al. developed such a hybrid platform by exploiting fluorinated ethylene propylene film as a TENG material and lithium titanate–based flexible LIB. When a TENG was exposed to a wind speed of 24.6 m s^{-1}, an output voltage of about 135 V along with an output current of 12 μA was generated. The TENG output can be boosted by a transformer and the rectified output voltage can charge the LIB within 55 s. This method offers a facile route for energy harvesting, storing, and longtime powering of wearable devices (552).

5.5 Thermoelectric battery/supercapacitor

Thermoelectric energy generators (TEGs) convert thermal energy into electrical energy without any gas emissions. The working principle behind the TEG is Seebeck effect, which is one of the thermoelectric effects, and defines that a voltage difference is produced between two dissimilar electrical conductors or semiconductors when there is a temperature difference between both the substances. The TEG device utilizes a temperature difference between the two ends of the device to govern the diffusion of charge carriers. The ability of a TEG to convert heat into electrical energy even with minimal temperature differences makes it appropriate for exploiting the body temperature and harvest energy for wearable applications. The Seebeck effect can be represented as equation (5.6):

$$\Delta V = \propto \Delta T \qquad (5.6)$$

where ΔV is the thermoelectric voltage, ΔT denotes the temperature gradient, and α indicates the Seebeck coefficient. According to the thermoelectric effect, the efficiency of thermoelectric devices is calculated by the figure of merit (ZT) of a thermoelectric material, which is a function of a number of transport coefficients as represented in equation (5.7):

$$ZT = \frac{\sigma S^2 T}{K_e + K_l} \qquad (5.7)$$

where σ indicates the electrical conductivity, S denotes the Seebeck coefficient, T is the mean operating temperature, and K represents the thermal conductivity. The subscripts e and l of K denote the electronic and lattice contributions, respectively.

Human body possesses very limited mechanical efficiency of 15–30% and majority of the consumed energy from food is released as heat. The normal body temperature is regulated at 37 °C and the heat can be used as a continuous energy source for TEGs. Harvesting energy from the body heat is usually sufficient to power many of the wearable electronics. Numerous semiconductor and ceramic materials have been reported so far for TEG application including PbTe (553), BiSbTe$_3$ (554), CoSb$_3$ (555), Mg$_2$Si (556), and SiGe (557). Several disadvantages are associated with inorganic thermoelectric materials such as lack of natural resources, toxicity, high cost, and complex manufacturing procedures. Introducing conducting polymers or conductive fillers with the thermoelectric material is appropriate for realizing high-performance TEGs for wearable applications (83). Apart from the ability to generate required power, the body-based energy generators should be compatible and comfortable to wear, without restricting the natural functions of the body. The use of bulky and rigid components may restrict the metabolism or involuntary movements of the body. Introducing flexibility and stretchability to TEGs is needed to realize skin conformable soft harvesters that can cover large areas of the human body and scale up the energy generation capacity.

Glass fabric is one of a good flexible substrate for fabricating wearable thermoelectric devices owing to the advantages such as excellent mechanical strength, corrosion-resistance, high tensile strength, high-temperature resistance, and low-cost (558). The low thickness of the glass fabric provides good flexibility, and its unique structure functions as a support for the thermoelectric material. Also, the glass fiber skeleton encapsulated inside the thick thermoelectric film serves as a thermal blocker and reduces the thermal conductivity. Therefore, the glass fiber enhances the performances of the TEG by reducing the body heat dissipation to the environment. Cho et al. developed a glass fabric–based flexible TEG using a screen printing technique. The method offers thin-film, lightweight TEGs with good flexibility, and the energy-harvesting performances are not affected by repeated bending cycles (559).

Silk has been regarded as "the queen of textiles" and possesses advantages like softness, large hygroscopicity, and superior skin affinity that make it an ideal substrate for developing wearable TEGs. Li et al. developed silk fabric–based TEG by depositing nanostructured Bi$_2$Te$_3$ and Sb$_2$Te$_3$ on both sides of the fabric to form n-type and p-type columns. The TEG can harvest waste heat produced from human body, and there is no considerable change in performance during 100 cycles of twisting and bending deformations (560). In order to establish continuous powering, Wang et al. developed a wearable TEG and integrated it with a flexible supercapacitor (561). TEGs are successfully merged with fashion sports equipment such as wrist guards, headbands, leg guards, or sports tights. The harvested energy from body temperature is used to charge the supercapacitor and the hybrid system provides stable power output, and can be used for powering human diagnostic devices.

5.6 Pyroelectric battery/supercapacitor

Pyroelectric effect is another methodology that utilizes thermal energy to harvest electrical energy, based on the change in spontaneous polarization in certain anisotropic solids due to the temperature fluctuation (562). Pyroelectricity is the ability of certain materials to produce a temporary voltage when they are exposed to heat or cold. The change in temperature slightly alters the positions of the atoms inside the crystal structure, resulting in changes in polarization of the materials. This polarization change establishes a voltage across the crystal. The device used to harvest pyroelectricity is termed as pyroelectric nanogenerator (PNG). At micro-/nanoscale, the pyroelectric materials are found to be more efficient for thermal energy conversion into electrical energy compared to thermoelectric materials, as they necessitate temporal temperature gradient (dT/dt) rather than a spatial temperature gradient (dT/dx), which is extremely complex to establish for nanostructured materials (563). PNG involves two working mechanisms such as primary and secondary pyroelectric effect. The primary pyroelectric effect involves the charge generation due to the change in polarization in the anisotropic crystals with the change in temperature when the dimensions of the material are constant. The additional secondary effect is due to the generation of piezoelectricity when there is a thermal expansion with the change in temperature. The pyroelectric coefficient, 'p', can be represented as equation (5.8):

$$e = \frac{d\rho}{dT} \tag{5.8}$$

where e denotes the pyroelectric coefficient, ρ indicates the spontaneous polarization, and T is the temperature. Table 5.2 represents the pyroelectric coefficients of various materials at room temperature (564–572). The electric current produced by the pyroelectric effect is given as equation (5.9):

$$I = \frac{dQ}{dt} = \mu e A \frac{dT}{dt} \tag{5.9}$$

where Q indicates the induced charge, μ is the absorption co-efficient of radiation, A represents the surface area, and dT/dt denotes the rate of temperature change. Pyroelectric materials possess high-temperature stability (~1,200 °C) that makes it suitable for high-temperature applications. Some methods are recently reported to convert stationary spatial gradients into transient temperature gradients that enable to develop hybrid energy harvesters with both thermo- and pyroelectric materials. PNGs do not require any specific maintenance and are independent of any moving components; thus, PNG-based energy harvesting is appropriate for wearable devices.

Significant temperature difference exists between the human body and the surrounding, particularly in the winter. The breathing process establishes time-dependent temperature variations nearby the nose and mouth. These temperature fluctuations can be exploited for PNG-based energy harvesting. Air pollution is a major threat during winter in many places of the world and the use of respirator is very common. Therefore, integrating PNG with the respirator provides close contact with the mouth and facilitates energy harvesting from the breath. The periodic temperature changes induce changes in polarization and generate potential between electrodes. Luo et al. developed a PVDF film–based PNG and integrated on an N95 respirator to scavenge pyroelectric energy from human respiration. Such a wearable device is appropriate for converting waste heat into electrical energy for self-powered healthcare monitoring devices, without adding any trouble for the wearer (573).

Table 5.2: Pyroelectric coefficient of different materials at room temperature.

Material	Point group	Primary coefficient $(m^{-2}K^{-1})$	Secondary coefficient $(m^{-2}K^{-1})$	Experimental value $(m^{-2}K^{-1})$	References
Triglycine sulfate	2	60	−330	−270	(564)
$Pb_5Ge_3O_{11}$	3	−110.5	+15.5	−95	(565)
$LiNbO_3$	3 m	−95.8	+12.8	−83	(566)
$BaTiO_3$	∞m	−260	+60	−200	(567)
$LiTaO_3$	3 m	−175	−1	−176	(568)
$PbZr_{0.95}Ti_{0.05}O_3$	∞m	−305.7	+37.7	−268	(569)
CdSe	6 mm	−2.94	−0.56	−3.5	(570)
ZnO	6 mm	−6.9	−2.5	−9.4	(571)
CdS	6 mm	−3.0	−1.0	−4.0	(572)

Pyroelectric materials are considered as the subclass of piezoelectric materials. Many of the piezoelectric materials possess pyroelectric behavior, in which electrical energy is produced with the application of a thermal gradient. The poly(vinylidenefluoride-co-trifluoroethylene) polymer [P(VDF-TrFE)] shows excellent piezoelectric behavior and pyroelectric properties with pyroelectric coefficient of around 200 $\mu C\ m^{-2}K^{-1}$. Kim et al. developed a stretchable hybrid nanogenerator using micro-patterned P(VDF-TrFE), PDMS-CNT composite, and graphene nanosheets. The hybrid device exhibits 30% stretchability and simultaneously generates robust piezoelectric and pyroelectric power outputs (574). Integrating multiple energy harvesters that can generate electricity from different sources provides stable extended power output. Wang et al. developed such a triboelectric–pyroelectric–piezoelectric hybrid device for highly efficient energy harvesting and self-powered sensing (575).

The continuous power generation from the pyroelectric devices depends on the continuous temporal changes in temperature. Introducing two different air sources having significant change in temperature can maintains the sequential temperature gradient. Also, the power density of PNG devices is very low and unable to power wearable sensors, continuously. The problems associated with the temperature fluctuations also adversely affect the energy-harvesting process. Integrating PNGs with capacitors or batteries can store the harvested energies, and thus they can function as a stable power sources. For instance, Chauhan et al. utilized pyroelectric ceramics for harvesting thermal energy from the hot and cold air sources (576). When integrated with a 4.7 μF capacitor, the hybrid device showed a maximum output voltage of 2.51 V with temperature disparity between 100 °C and 65 °C. Besides, the PNG can charge a battery of 7 mAh capacity to 1.21 V within 55 min when it is exposed to temporal heating and cooling procedures with a frequency of 0.4 Hz.

5.7 Biofuel cell battery/supercapacitor

A biofuel cell uses living organisms to harvest electrical energy. There are two types of biofuel cells: microbial fuel cell and enzymatic biofuel cell. The microbial fuel cell converts chemical energy into electrical energy using bacteria and mimicking other bacterial interactions exist in nature. But the enzymatic biofuel cell is a kind of fuel cell that uses enzymes as electrocatalysts to oxidize biomolecules as its fuel. Microbial fuel cell possesses multiple oxido-reductase enzymes that provide complete oxidation of different fuels. Some limitations restrict its use for wearable application; for instance, the active sites of enzymes are buried well inside the cell walls of the microbes. This impedes the effective fuel accessibility to enzymes and reduces the electron transfer rates, thus resulting in poor power density. In addition, the cytotoxicity of microbial fuel cells restricts them from being chosen as a wearable power energy source (577, 578).

Enzymatic biofuel cells are promising biotic power sources for various modes of implantable, invasive, minimally invasive, and non-invasive applications. A typical enzymatic biofuel cell consists of an enzyme modified bioanode and biocathode. Hybrid enzymatic biofuel cells having enzyme-based bioanode and noble metal catalyst functionalized cathode are also exploited as a wearable power source (579). In anode, the oxidation of fuels like glucose, lactate, pyruvate, cellulose, and ethanol happens by highly selective enzymes, and electrons are generated. The cathode reaction typically involves oxygen reduction reaction either by utilizing laccase or bilirubin oxidase enzymes or by exploiting noble metal–based catalysts like Pt, Pd, or Ru. Although the metal catalysts provide high energy density, the enzyme-based biocathodes present advantages like low cost, large open circuit potential, and appreciable selectivity.

Human sweat is an abundant source of metabolites that can be utilized as fuels for enzymatic biofuel cells. Lactate-based biofuel cells are getting considerable attraction as the lactate production is related to the fitness level and the oxygenation of muscles, and its concentration in sweat is increased with respect to the exercise intensities. The average concentration of sweat lactate is 14 mM and it can be increase to as high as 50 mM, and thus the lactate biofuel cells can harvest sufficient electrical energy to power wearable health- and fitness-monitoring devices. Ideally, wearable biofuel cells should be thin, soft, compact, and stretchable, in order to achieve conformal integration with the human skin or other platforms like temporary tattoos or textiles. Wang et al. developed a soft, deterministically stretchable electronic skin like biofuel cell that can harvest a high power density of 1 mW cm^{-2} from sweat lactate during exercise activities (421). The lithographically patterned island-bridge configuration provides stretchability, and the use of densely packed screen-printed thick film offers high power density. The energy produced from this biofuel cell was able to power light-emitting diodes and even a Bluetooth low-energy radio.

Majority of the wearable devices used for healthcare monitoring, electronic displays, fitness monitoring, motion monitoring, thermal controlling, and many more necessitate continuous energy over long periods of time. The energy produced from the biofuel cells is highly limited to the availability of fuels. For instance, the sweat-based lactate biofuel cell works well only during high-endurance activities and are unable to act as a power source in the absence of sufficient lactate concentrations. When integrated with wearable energy storage devices, the harvested energy from the biofuel cell can be periodically stored as and when the fuel is available. Wang et al. developed such a hybrid powering system by combining textile-based stretchable biofuel cell and a supercapacitor (Figure 5.3a). The biofuel cell attached with the textile can access sweat from the skin surface, and the harvested energy can be transferred to the supercapacitor integrated on the reverse side. The self-charging hybrid system can maintain its performances under bending, twisting, and stretching deformations (Figure 5.3b). Such combination facilitates stabilized output even if there are fluctuations in sweat lactate levels, and this is a vital step toward the development of wearable self-powered electronics (580). Sweat-activated biocompatible batteries are another emerging technology which is different from biofuel cells, but they can be integrated inside a soft, microfluidic platform and exploit the advantages of collected sweat components for better battery performances (581).

5.8 Power management

The output from many of the energy harvesters or storage devices is not enough or suitable to directly power the wearable electronics. The energy output from triboelectric, piezoelectric, thermoelectric, and pyroelectric devices are alternating current (AC),

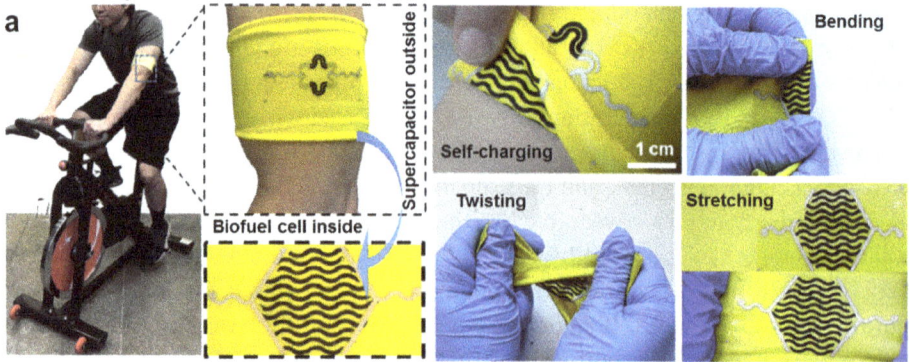

Figure 5.3: (a) Photographs showing hybrid supercapacitor – Biofuel cell self-powered system on a textile substrate, applied to the arm of a volunteer. The supercapacitor and biofuel cell are printed outside and inside the arm band, respectively. (b) Images showing the stretchable hybrid device printed on a wearable wristband, under conditions of bending, twisting, and stretching deformations (reproduced with permission from ref. (580)).

which is an electric current that periodically reverses its direction and changes magnitude continuously with time. But the common power source for wearable devices is direct current (DC) that flows only in one direction. Hence, these energy harvesters need to be interfaced with a rectifier which can convert AC to DC and power the external devices. The ultralow-power output from the energy harvesters is another bottleneck for efficient charging or powering. The use of step-up transformers can address this problem by increasing low AC voltages at high current. A transformer is a passive electrical device that transfers electrical energy from one circuit to another through the process of electromagnetic induction. The equivalent circuit used for battery charging by the TENG with the help of a transformer and a rectifier is represented in Figure 5.4a (428). When the output voltage from the energy harvesters or the hybrid devices is lower than the operational voltage of the wearable device, the use of DC–DC boost converter can increase the output voltage, though the net power remains constant. For efficient self-charging system, the low-power TENG harvesters are connected with AC–DC buck converter, rectifier, and conventional DC–DC buck converter (Figure 5.4b). Such integration enables maximizing the efficiency and providing sustainable power for wearable/portable electronics or sensors (429). In general, the power management circuit should be capable of boosting the current, lowering the impedance, and supplying consistent power to a variety of electronics.

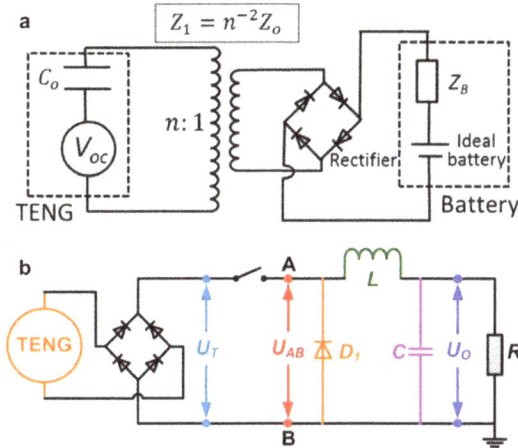

Figure 5.4: (a) The equivalent circuit diagram of battery charging by the TENG with the aid of a transformer and a rectifier (reproduced with permission from ref. (428)). (b) The schematic circuit pattern of AC–DC buck conversion by coupling TENG, rectifier and classical DC–DC buck converter (reproduced with permission from ref. (429)).

5.9 Summary and outlook

The existing wearable energy storage devices necessitate frequent charging as their energy densities are insufficient to meet the power requirement for continuous long-term operation. Combining the energy-harvesting devices with the energy storage systems offers high energy and power densities for robust powering of wearable electronics. This chapter discussed the recent advances in such hybrid systems which can function as independent self-powering platform. Energy harvesters like photovoltaic, piezoelectric, triboelectric, thermoelectric, and pyroelectric devices are suitable for generating energy from sun light, pressure, friction, vibration, wind, heat, and many more. This realizes harvesting energy from normal body motion during daily activities or from body temperature and storing it into supercapacitors or batteries. Also, biofuel cells convert the chemical energy from human body fluids into electrical energy by utilizing specific enzyme-based catalysts. The energy output from the harvesters can be converted or boosted up by interfacing with rectifiers, amplifiers, or DC–DC converters. These hybrid platforms avoid the energy buffering issues, separate charging requirement, and maintains stable powering. The mechanical properties of these hybrid devices are significant as all the wearable platforms mandate soft, flexible, and stretchable counterparts.

The book elaborates the burgeoning wearable energy storage devices useful for powering various wearable electronics. The healthcare- or fitness-tracking devices need stable power supply for continuous biomarker evaluation, health monitoring, data collection, and wireless data transduction. The mechanical properties of all the wearable devices should match with that of the substrates to avoid device fatigue and failure. Developing soft, flexible, stretchable, and self-healing devices is really crucial for wearable applications. The recent advances in the development of wearable supercapacitors and batteries are summarized, along with the synthesis and applications of stretchable and self-healing gel electrolytes. Furthermore, the amalgamation of energy-harvesting and storage systems for self-powering application is detailed. These wearable devices are significant for developing next-generation healthcare trackers for point-of-care analysis, and also for auto-response drug delivery applications.

References

1. Gupta S, Navaraj WT, Lorenzelli L, Dahiya R. Ultra-thin chips for high-performance flexible electronics. npj Flexible Electronics. 2018;2(1):1–17.
2. Gates BD. Flexible electronics. Science. 2009;323(5921):1566–67.
3. Wong WS, Salleo A. Flexible Electronics: Materials and Applications: Springer Science & Business Media; 2009.
4. Nathan A, Ahnood A, Cole MT, Lee S, Suzuki Y, Hiralal P, et al. Flexible electronics: the next ubiquitous platform. Proceedings of the IEEE. 2012;100(Special Centennial Issue):1486–517.
5. Chen X, Rogers JA, Lacour SP, Hu W, Kim D-H. Materials chemistry in flexible electronics. Chemical Society Reviews. 2019;48(6):1431–33.
6. Ray TR, Choi J, Bandodkar AJ, Krishnan S, Gutruf P, Tian L, et al. Bio-integrated wearable systems: a comprehensive review. Chemical Reviews. 2019;119(8):5461–533.
7. Kim SH, Moon J-H, Kim JH, Jeong SM, Lee S-H. Flexible, stretchable and implantable PDMS encapsulated cable for implantable medical device. Biomedical Engineering Letters. 2011;1(3):199–203.
8. Zhao X, Hua Q, Yu R, Zhang Y, Pan C. Flexible, stretchable and wearable multifunctional sensor array as artificial electronic skin for static and dynamic strain mapping. Advanced Electronic Materials. 2015;1(7):1500142.
9. Fujikake H, Sato H, editors. Flexible ferroelectric liquid crystal devices for roll-up displays. Liquid Crystal Materials, Devices, and Applications X and Projection Displays X; 2004: International Society for Optics and Photonics.
10. Mohan AV, Windmiller JR, Mishra RK, Wang J. Continuous minimally-invasive alcohol monitoring using microneedle sensor arrays. Biosensors and Bioelectronics. 2017;91:574–79.
11. Mishra RK, Mohan AV, Soto F, Chrostowski R, Wang J. A microneedle biosensor for minimally-invasive transdermal detection of nerve agents. Analyst. 2017;142(6):918–24.
12. Bandodkar AJ, Imani S, Nunez-Flores R, Kumar R, Wang C, Mohan AV, et al. Re-usable electrochemical glucose sensors integrated into a smartphone platform. Biosensors and Bioelectronics. 2018;101:181–87.
13. Iqbal SM, Mahgoub I, Du E, Leavitt MA, Asghar W. Advances in healthcare wearable devices. npj Flexible Electronics. 2021;5(1):1–14.
14. Lee H, Song C, Hong YS, Kim MS, Cho HR, Kang T, et al. Wearable/disposable sweat-based glucose monitoring device with multistage transdermal drug delivery module. Science Advances. 2017;3(3):e1601314.
15. Imani S, Bandodkar AJ, Mohan AV, Kumar R, Yu S, Wang J, et al. A wearable chemical–electrophysiological hybrid biosensing system for real-time health and fitness monitoring. Nature Communications. 2016;7(1):1–7.
16. Wang Y, Wang L, Yang T, Li X, Zang X, Zhu M, et al. Wearable and highly sensitive graphene strain sensors for human motion monitoring. Advanced Functional Materials. 2014;24(29):4666–70.
17. Sung M, Marci C, Pentland A. Wearable feedback systems for rehabilitation. Journal of Neuroengineering and Rehabilitation. 2005;2(1):1–12.
18. Gao W, Ota H, Kiriya D, Takei K, Javey A. Flexible electronics toward wearable sensing. Accounts of Chemical Research. 2019;52(3):523–33.
19. Bandodkar AJ, Jeerapan I, Wang J. Wearable chemical sensors: present challenges and future prospects. ACS Sensors. 2016;1(5):464–82.
20. Bandodkar AJ, Wang J. Non-invasive wearable electrochemical sensors: a review. Trends in Biotechnology. 2014;32(7):363–71.
21. Cima MJ. Next-generation wearable electronics. Nature Biotechnology. 2014;32(7):642–43.

https://doi.org/10.1515/9781501521287-006

22. Koo J, Kim SB, Choi YS, Xie Z, Bandodkar AJ, Khalifeh J, et al. Wirelessly controlled, bioresorbable drug delivery device with active valves that exploit electrochemically triggered crevice corrosion. Science Advances. 2020;6(35):eabb1093.

23. Gillinov S, Etiwy M, Wang R, Blackburn G, Phelan D, Gillinov AM, et al. Variable accuracy of wearable heart rate monitors during aerobic exercise. Medicine and Science in Sports and Exercise. 2017;49(8):1697–703.

24. Arakawa T. Recent research and developing trends of wearable sensors for detecting blood pressure. Sensors. 2018;18(9):2772.

25. Jeong J, Jang Y, Lee I, Shin S, Kim S, editors. Wearable respiratory rate monitoring using piezo-resistive fabric sensor. World Congress on Medical Physics and Biomedical Engineering, september 7–12, 2009, munich, germany; 2009: Springer.

26. Davies HJ, Williams I, Peters NS, Mandic DP. In-Ear SpO2: a Tool for Wearable, Unobtrusive Monitoring of Core Blood Oxygen Saturation. Sensors. 2020;20(17):4879.

27. Nakata S, Arie T, Akita S, Takei K. Wearable, flexible, and multifunctional healthcare device with an ISFET chemical sensor for simultaneous sweat pH and skin temperature monitoring. ACS sensors. 2017;2(3):443–48.

28. Montesinos L, Castaldo R, Pecchia L. Wearable inertial sensors for fall risk assessment and prediction in older adults: a systematic review and meta-analysis. IEEE Transactions on Neural Systems and Rehabilitation Engineering. 2018;26(3):573–82.

29. Kim SB, Lee K, Raj MS, Lee B, Reeder JT, Koo J, et al. Soft, Skin-Interfaced Microfluidic Systems with Wireless, Battery-Free Electronics for Digital, Real-Time Tracking of Sweat Loss and Electrolyte Composition. Small. 2018;14(45):1802876.

30. Vinoth R, Nakagawa T, Mathiyarasu J, Mohan AV. Fully Printed Wearable Microfluidic Devices for High-Throughput Sweat Sampling and Multiplexed Electrochemical Analysis. ACS Sensors. 2021;6(3):1174–86.

31. Mohan AV, Rajendran V, Mishra RK, Jayaraman M. Recent advances and perspectives in sweat based wearable electrochemical sensors. TrAC Trends in Analytical Chemistry. 2020:116024.

32. García-Carmona L, Martín A, Sempionatto JR, Moreto JR, González MaC, Wang J, et al. Pacifier biosensor: toward noninvasive saliva biomarker monitoring. Analytical Chemistry. 2019;91 (21):13883–91.

33. Zhang J, Liu J, Su H, Sun F, Lu Z, Su A. A wearable self-powered biosensor system integrated with diaper for detecting the urine glucose of diabetic patients. Sensors and Actuators B: Chemical. 2021:130046.

34. Zhou J, Dong T. Design of a wearable device for real-time screening of urinary tract infection and kidney disease based on smartphone. Analyst. 2018;143(12):2812–18.

35. Heikenfeld J, Jajack A, Rogers J, Gutruf P, Tian L, Pan T, et al. Wearable sensors: modalities, challenges, and prospects. Lab on a Chip. 2018;18(2):217–48.

36. Kim J, Campbell AS, de Ávila B, Wang J. Wearable biosensors for healthcare monitoring. Nature Biotechnology. 2019;37(4):389–406.

37. Yang Y, Gao W. Wearable and flexible electronics for continuous molecular monitoring. Chemical Society Reviews. 2019;48(6):1465–91.

38. Wang L, Chen D, Jiang K, Shen G. New insights and perspectives into biological materials for flexible electronics. Chemical Society Reviews. 2017;46(22):6764–815.

39. Nambiar SR, Mohan AV. Metal Nanoparticles-based Disposable Sensors. Disposable Electrochemical Sensors for Healthcare Monitoring2021. p. 170–203.

40. Vinoth R, Nakagawa T, Mathiyarasu J, Mohan AV. Fully Printed Wearable Microfluidic Devices for High-Throughput Sweat Sampling and Multiplexed Electrochemical Analysis. ACS Sensors. 2021.

41. Bandodkar AJ, Jeang WJ, Ghaffari R, Rogers JA. Wearable sensors for biochemical sweat analysis. Annual Review of Analytical Chemistry. 2019;12:1–22.

42. Kim J, Jeerapan I, Imani S, Cho TN, Bandodkar A, Cinti S, et al. Noninvasive alcohol monitoring using a wearable tattoo-based iontophoretic-biosensing system. Acs Sensors. 2016;1(8):1011–19.

43. Kim J, Sempionatto JR, Imani S, Hartel MC, Barfidokht A, Tang G, et al. Simultaneous monitoring of sweat and interstitial fluid using a single wearable biosensor platform. Advanced Science. 2018;5(10):1800880.

44. Tai LC, Gao W, Chao M, Bariya M, Ngo QP, Shahpar Z, et al. Methylxanthine drug monitoring with wearable sweat sensors. Advanced Materials. 2018;30(23):1707442.

45. Lin S, Wang B, Yu W, Castillo K, Hoffman C, Cheng X, et al. Design framework and sensing system for noninvasive wearable electroactive drug monitoring. ACS Sensors. 2020;5(1):265–73.

46. Sweilam MN, Cordery SF, Totti S, Velliou EG, Campagnolo P, Varcoe JR, et al. Textile-based non-invasive lithium drug monitoring: a proof-of-concept study for wearable sensing. Biosensors and Bioelectronics. 2020;150:111897.

47. Teymourian H, Parrilla M, Sempionatto JR, Montiel NF, Barfidokht A, Van Echelpoel R, et al. Wearable electrochemical sensors for the monitoring and screening of drugs. ACS Sensors. 2020;5(9):2679–700.

48. Carreiro S, Smelson D, Ranney M, Horvath KJ, Picard RW, Boudreaux ED, et al. Real-time mobile detection of drug use with wearable biosensors: a pilot study. Journal of Medical Toxicology. 2015;11(1):73–79.

49. Mahmud MS, Fang H, Carreiro S, Wang H, Boyer EW. Wearables technology for drug abuse detection: a survey of recent advancement. Smart Health. 2019;13:100062.

50. Torrente-Rodríguez RM, Tu J, Yang Y, Min J, Wang M, Song Y, et al. Investigation of cortisol dynamics in human sweat using a graphene-based wireless mHealth system. Matter. 2020;2(4):921–37.

51. Ates HC, Yetisen AK, Güder F, Dincer C. Wearable devices for the detection of COVID-19. Nature Electronics. 2021;4(1):13–14.

52. Quer G, Radin JM, Gadaleta M, Baca-Motes K, Ariniello L, Ramos E, et al. Wearable sensor data and self-reported symptoms for COVID-19 detection. Nature Medicine. 2021;27(1):73–77.

53. Bandodkar AJ, Jia W, Wang J. Tattoo-based wearable electrochemical devices: a review. Electroanalysis. 2015;27(3):562–72.

54. Li L, Kim S, Wang W, Vijayakumar M, Nie Z, Chen B, et al. A stable vanadium redox-flow battery with high energy density XE "energy density" for large-scale energy storage. Advanced Energy Materials. 2011;1(3):394–400.

55. Chen Q, Shen Y, Zhang S, Zhang Q. Polymer-based dielectrics with high energy storage density. Annual Review of Materials Research. 2015;45:433–58.

56. Ibrahim H, Ilinca A, Perron J. Energy storage systems – Characteristics and comparisons. Renewable and Sustainable Energy Reviews. 2008;12(5):1221–50.

57. Ribeiro PF, Johnson BK, Crow ML, Arsoy A, Liu Y. Energy storage systems for advanced power applications. Proceedings of the IEEE. 2001;89(12):1744–56.

58. Windmiller JR, Wang J. Wearable electrochemical sensors and biosensors: a review. Electroanalysis. 2013;25(1):29–46.

59. Akindoyo JO, Beg M, Ghazali S, Islam M, Jeyaratnam N, Yuvaraj A. Polyurethane types, synthesis and applications–a review. RSC Advances. 2016;6(115):114453–82.

60. Wang B, Facchetti A. Mechanically flexible conductors for stretchable and wearable e-skin and e-textile devices. Advanced Materials. 2019;31(28):1901408.

61. Pinchuk L, Riss I, Batlle JF, Kato YP, Martin JB, Arrieta E, et al. The development of a micro-shunt made from poly (styrene-block-isobutylene-block-styrene) to treat glaucoma. Journal of Biomedical Materials Research Part B: Applied Biomaterials. 2017;105(1):211–21.

62. Garalde RA, Thipmanee R, Jariyasakoolroj P, Sane A. The effects of blend ratio and storage time on thermoplastic starch/poly (butylene adipate-co-terephthalate) films. Heliyon. 2019;5 (3):e01251.

63. Shintake J, Sonar H, Piskarev E, Paik J, Floreano D, editors. Soft pneumatic gelatin actuator for edible robotics. 2017 IEEE/RSJ International Conference on Intelligent Robots and Systems (IROS); 2017: IEEE.

64. Park JM, Son J, An HJ, Kim JH, Wu H-G, Kim J-i. Bio-compatible patient-specific elastic bolus for clinical implementation. Physics in Medicine & Biology. 2019;64(10):105006.

65. Abd Kati F. Physical properties of silicone elastomers. Review. 2019.

66. Gorkem C, Sengoz B. Predicting stripping and moisture induced damage of asphalt concrete prepared with polymer modified bitumen and hydrated lime. Construction and Building Materials. 2009;23(6):2227–36.

67. Eissa M, Botros S, Moustafa A. Effect of triblock copolymers on homogeneity, mechanical properties and swelling behavior of IIR/SBR rubber blends. Polymer Bulletin. 2017;74 (2):393–412.

68. Denac M, Musil V, Šmit I. Polypropylene/ talc/ SEBS (SEBS-g-MA) composites. Part 2. Mechanical properties. Composites Part A: Applied Science and Manufacturing. 2005;36 (9):1282–90.

69. Miao R, Liu B, Zhu Z, Liu Y, Li J, Wang X, et al. PVDF-HFP-based porous polymer electrolyte membranes for lithium-ion batteries. Journal of Power Sources. 2008;184(2):420–26.

70. Wu YT, Stewart MA. Ethylene acrylic elastomers. Encyclopedia of Polymer Science and Technology. 2002.

71. Lee B-C, Jeong WB, Lee KD. Phase Behavior of Ethylene-co-Vinyl Acetate and Alkyl Acrylate Copolymer in High-Pressure Dimethyl Ether. Chemical Engineering & Technology. 2017;40 (12):2221–29.

72. Yoon Y, Samanta K, Lee H, Lee K, Tiwari AP, Lee J, et al. Highly stretchable and conductive silver nanoparticle embedded graphene flake electrode prepared by in situ dual reduction reaction. Scientific Reports. 2015;5(1):1–10.

73. Sekitani T, Noguchi Y, Hata K, Fukushima T, Aida T, Someya T. A rubberlike stretchable active matrix using elastic conductors. Science. 2008;321(5895):1468–72.

74. Akter T, Kim WS. Reversibly stretchable transparent conductive coatings of spray-deposited silver nanowires. ACS Applied Materials & Interfaces. 2012;4(4):1855–59.

75. Hu W, Niu X, Li L, Yun S, Yu Z, Pei Q. Intrinsically stretchable transparent electrodes based on silver-nanowire–crosslinked-polyacrylate composites. Nanotechnology. 2012;23(34):344002.

76. Cheng Y, Wang S, Wang R, Sun J, Gao L. Copper nanowire based transparent conductive films with high stability and superior stretchability. Journal of Materials Chemistry C. 2014;2 (27):5309–16.

77. Hu W, Wang R, Lu Y, Pei Q. An elastomeric transparent composite electrode based on copper nanowires and polyurethane. Journal of Materials Chemistry C. 2014;2(7):1298–305.

78. Yao S, Zhu Y. Wearable multifunctional sensors using printed stretchable conductors made of silver nanowires. Nanoscale. 2014;6(4):2345–52.

79. Xu F, Zhu Y. Highly conductive and stretchable silver nanowire conductors. Advanced Materials. 2012;24(37):5117–22.

80. Park S-M, Jang N-S, Ha S-H, Kim KH, Jeong D-W, Kim J, et al. Metal nanowire percolation micro-grids embedded in elastomers for stretchable and transparent conductors. Journal of Materials Chemistry C. 2015;3(31):8241–47.

81. Kim D-H, Yu K-C, Kim Y, Kim J-W. Highly stretchable and mechanically stable transparent electrode based on composite of silver nanowires and polyurethane–urea. ACS Applied Materials & Interfaces. 2015;7(28):15214–22.

82. Cheng T, Zhang Y-Z, Lai W-Y, Chen Y, Zeng W-J, Huang W. High-performance stretchable transparent electrodes based on silver nanowires synthesized via an eco-friendly halogen-free method. Journal of Materials Chemistry C. 2014;2(48):10369–76.

83. Li P, Sun K, Ouyang J. Stretchable and conductive polymer films prepared by solution blending. ACS Applied Materials & Interfaces. 2015;7(33):18415–23.

84. An BW, Hyun BG, Kim S-Y, Kim M, Lee M-S, Lee K, et al. Stretchable and transparent electrodes using hybrid structures of graphene–metal nanotrough networks with high performances and ultimate uniformity. Nano Letters. 2014;14(11):6322–28.

85. Lee M-S, Lee K, Kim S-Y, Lee H, Park J, Choi K-H, et al. High-performance, transparent, and stretchable electrodes using graphene–metal nanowire XE "metal nanowire" hybrid structures. Nano Letters. 2013;13(6):2814–21.

86. Matsuhisa N, Kaltenbrunner M, Yokota T, Jinno H, Kuribara K, Sekitani T, et al. Printable elastic conductors with a high conductivity for electronic textile applications. Nature Communications. 2015;6(1):1–11.

87. Yan C, Wang J, Kang W, Cui M, Wang X, Foo CY, et al. Highly stretchable piezoresistive graphene–nanocellulose nanopaper for strain sensors. Advanced Materials. 2014;26 (13):2022–27.

88. Tang Y, Zhao Z, Hu H, Liu Y, Wang X, Zhou S, et al. Highly stretchable and ultrasensitive strain sensor based on reduced graphene oxide microtubes–elastomer composite. ACS Applied Materials & Interfaces. 2015;7(49):27432–39.

89. Jeong YR, Park H, Jin SW, Hong SY, Lee SS, Ha JS. Highly stretchable and sensitive strain sensors using fragmentized graphene foam. Advanced Functional Materials. 2015;25 (27):4228–36.

90. Roh E, Hwang B-U, Kim D, Kim B-Y, Lee N-E. Stretchable, transparent, ultrasensitive, and patchable strain sensor for human–machine interfaces comprising a nanohybrid of carbon nanotubes and conductive elastomers. ACS Nano. 2015;9(6):6252–61.

91. Lee S, Shin S, Lee S, Seo J, Lee J, Son S, et al. Ag nanowire reinforced highly stretchable conductive fibers for wearable electronics. Advanced Functional Materials. 2015;25(21):3114–21.

92. Hwang B-U, Lee J-H, Trung TQ, Roh E, Kim D-I, Kim S-W, et al. Transparent stretchable self-powered patchable sensor platform with ultrasensitive recognition of human activities. ACS Nano. 2015;9(9):8801–10.

93. Amjadi M, Pichitpajongkit A, Lee S, Ryu S, Park I. Highly stretchable and sensitive strain sensor based on silver nanowire–elastomer nanocomposite. ACS Nano. 2014;8(5):5154–63.

94. Chen M, Zhang L, Duan S, Jing S, Jiang H, Li C. Highly stretchable conductors integrated with a conductive carbon nanotube/graphene network and 3D porous poly (dimethylsiloxane). Advanced Functional Materials. 2014;24(47):7548–56.

95. Hansen TS, West K, Hassager O, Larsen NB. Highly stretchable and conductive polymer material made from poly (3, 4-ethylenedioxythiophene) and polyurethane elastomers. Advanced Functional Materials. 2007;17(16):3069–73.

96. Bhagavatheswaran ES, Parsekar M, Das A, Le HH, Wiessner S, Stöckelhuber KW, et al. Construction of an interconnected nanostructured carbon black network: development of highly stretchable and robust elastomeric conductors. The Journal of Physical Chemistry C. 2015;119(37):21723–31.

97. Ki H, Jang J, Jo Y, Kim D-Y, Chee S-S, Oh B-Y, et al. Chemically driven, water-soluble composites of carbon nanotubes and silver nanoparticles as stretchable conductors. ACS Macro Letters. 2015;4(7):769–73.

98. Lee YY, Kang HY, Gwon SH, Choi GM, Lim SM, Sun JY, et al. A strain-insensitive stretchable electronic conductor: PEDOT: PSS/acrylamide organogels. Advanced Materials. 2016;28 (8):1636–43.

99. Jiang S, Zhang H, Song S, Ma Y, Li J, Lee GH, et al. Highly stretchable conductive fibers from few-walled carbon nanotubes coated on poly (m-phenylene isophthalamide) polymer core/ shell structures. ACS Nano. 2015;9(10):10252–57.

100. Oh JY, Kim S, Baik HK, Jeong U. Conducting polymer dough for deformable electronics. Advanced Materials. 2016;28(22):4455–61.

101. Jeong GS, Baek D-H, Jung HC, Song JH, Moon JH, Hong SW, et al. Solderable and electroplatable flexible electronic circuit on a porous stretchable elastomer. Nature Communications. 2012;3(1):1–8.

102. Larmagnac A, Eggenberger S, Janossy H, Vörös J. Stretchable electronics based on Ag-PDMS composites. Scientific Reports. 2014;4(1):1–7.

103. Kim S, Byun J, Choi S, Kim D, Kim T, Chung S, et al. Negatively strain-dependent electrical resistance of magnetically arranged nickel composites: application to highly stretchable electrodes and stretchable lighting devices. Advanced Materials. 2014;26(19):3094–99.

104. Hu M, Cai X, Guo Q, Bian B, Zhang T, Yang J. Direct pen writing of adhesive particle-free ultrahigh silver salt-loaded composite ink for stretchable circuits. ACS Nano. 2016;10 (1):396–404.

105. Sekitani T, Nakajima H, Maeda H, Fukushima T, Aida T, Hata K, et al. Stretchable active-matrix organic light-emitting diode display using printable elastic conductors. Nature Materials. 2009;8(6):494–99.

106. Choi WM, Song J, Khang D-Y, Jiang H, Huang YY, Rogers JA. Biaxially stretchable "wavy" silicon nanomembranes. Nano Letters. 2007;7(6):1655–63.

107. Kim D-H, Song J, Choi WM, Kim H-S, Kim R-H, Liu Z, et al. Materials and noncoplanar mesh designs for integrated circuits with linear elastic responses to extreme mechanical deformations. Proceedings of the National Academy of Sciences. 2008;105(48):18675–80.

108. Fan JA, Yeo W-H, Su Y, Hattori Y, Lee W, Jung S-Y, et al. Fractal design concepts for stretchable electronics. Nature Communications. 2014;5(1):1–8.

109. Jang K-I, Li K, Chung HU, Xu S, Jung HN, Yang Y, et al. Self-assembled three dimensional network designs for soft electronics. Nature Communications. 2017;8(1):1–10.

110. Someya T, Kato Y, Sekitani T, Iba S, Noguchi Y, Murase Y, et al. Conformable, flexible, large-area networks of pressure and thermal sensors with organic transistor active matrixes. Proceedings of the National Academy of Sciences. 2005;102(35):12321–25.

111. Jang HY, Lee S-K, Cho SH, Ahn J-H, Park S. Fabrication of metallic nanomesh: Pt nano-mesh as a proof of concept for stretchable and transparent electrodes. Chemistry of Materials. 2013;25(17):3535–38.

112. Kim H-J, Son C, Ziaie B. A multiaxial stretchable interconnect using liquid-alloy-filled elastomeric microchannels. Applied Physics Letters. 2008;92(1):011904.

113. Yang Y, Sun N, Wen Z, Cheng P, Zheng H, Shao H, et al. Liquid-metal-based super-stretchable and structure-designable triboelectric nanogenerator for wearable electronics. ACS Nano. 2018;12(2):2027–34.

114. Xie Y, Liu Y, Zhao Y, Tsang YH, Lau SP, Huang H, et al. Stretchable all-solid-state supercapacitor with wavy shaped polyaniline/graphene electrode. Journal of Materials Chemistry A. 2014;2(24):9142–49.

115. Kim KM, Lee JA, Sim HJ, Kim K-A, Jalili R, Spinks GM, et al. Shape-engineerable composite fibers and their supercapacitor application. Nanoscale. 2016;8(4):1910–14.

116. Liu W, Chen J, Chen Z, Liu K, Zhou G, Sun Y, et al. Stretchable lithium-ion batteries enabled by device-scaled wavy structure and elastic-sticky separator. Advanced Energy Materials. 2017;7 (21):1701076.
117. Zhang W, Shen Z, Li S, Fan L, Wang X, Chen F, et al. Engineering wavy-nanostructured anode interphases with fast ion transfer kinetics: toward practical Li-Metal full batteries. Advanced Functional Materials. 2020;30(39):2003800.
118. Li R, Li M, Su Y, Song J, Ni X. An analytical mechanics model for the island-bridge structure of stretchable electronics. Soft Matter. 2013;9(35):8476–82.
119. Song J, Huang Y, Xiao J, Wang S, Hwang K-C, Ko H, et al. Mechanics of noncoplanar mesh design for stretchable electronic circuits. Journal of Applied Physics. 2009;105(12):123516.
120. Su Y, Wu J, Fan Z, Hwang K-C, Song J, Huang Y, et al. Postbuckling analysis and its application to stretchable electronics. Journal of the Mechanics and Physics of Solids. 2012;60(3):487–508.
121. Chen C, Tao W, Su Y, Wu J, Song J. Lateral buckling of interconnects in a noncoplanar mesh design for stretchable electronics. Journal of Applied Mechanics. 2013;80(4).
122. Conway BE. Electrochemical Supercapacitors: scientific Fundamentals and Technological Applications: Springer Science & Business Media; 2013.
123. Wang G, Zhang L, Zhang J. A review of electrode materials for electrochemical supercapacitors. Chemical Society Reviews. 2012;41(2):797–828.
124. Vivekchand S, Rout CS, Subrahmanyam K, Govindaraj A, Rao C. Graphene-based electrochemical supercapacitors. Journal of Chemical Sciences. 2008;120(1):9–13.
125. Lang X, Hirata A, Fujita T, Chen M. Nanoporous metal/oxide hybrid electrodes for electrochemical supercapacitors. Nature Nanotechnology. 2011;6(4):232–36.
126. Yan J, Wang Q, Wei T, Fan Z. Recent advances in design and fabrication of electrochemical supercapacitors with high energy densities. Advanced Energy Materials. 2014;4(4):1300816.
127. Arbizzani C, Mastragostino M, Soavi F. New trends in electrochemical supercapacitors. Journal of Power Sources. 2001; 100 (1–2): 164–70.
128. Hu C-C, Chang K-H, Lin M-C, Wu Y-T. Design and tailoring of the nanotubular arrayed architecture of hydrous RuO2 for next generation supercapacitors. Nano Letters. 2006;6 (12):2690–95.
129. Conway BE, Birss V, Wojtowicz J. The role and utilization of pseudocapacitance for energy storage by supercapacitors. Journal of Power Sources. 1997; 66 (1–2): 1–14.
130. Chen S, Zhu J, Wu X, Han Q, Wang X. Graphene oxide– MnO2 nanocomposites for supercapacitors. ACS Nano. 2010;4(5):2822–30.
131. Xia H, Meng YS, Yuan G, Cui C, Lu L. A symmetric RuO2/RuO2 supercapacitor operating at 1.6 V by using a neutral aqueous electrolyte. Electrochemical and Solid State Letters. 2012;15(4): A60.
132. Korkmaz S, Tezel FM, Kariper İ. Synthesis and characterization of GO/IrO2 thin film supercapacitor. Journal of Alloys and Compounds. 2018;754:14–25.
133. Du X, Wang C, Chen M, Jiao Y, Wang J. Electrochemical performances of nanoparticle Fe3O4/ activated carbon supercapacitor using KOH electrolyte solution. The Journal of Physical Chemistry C. 2009;113(6):2643–46.
134. Nandi D, Mohan VB, Bhowmick AK, Bhattacharyya D. Metal/metal oxide decorated graphene synthesis and application as supercapacitor: a review. Journal of Materials Science. 2020;55 (15):6375–400.
135. Barik R, Ingole PP. Challenges and prospects of metal sulfide materials for supercapacitors. Current Opinion in Electrochemistry. 2020.

136. Alvi F, Ram MK, Basnayaka PA, Stefanakos E, Goswami Y, Kumar A. Graphene–polyethylenedioxythiophene conducting polymer nanocomposite based supercapacitor. Electrochimica Acta. 2011;56(25):9406–12.

137. Zhao Y, Zhang Z, Ren Y, Ran W, Chen X, Wu J, et al. Vapor deposition polymerization of aniline on 3D hierarchical porous carbon with enhanced cycling stability as supercapacitor electrode. Journal of Power Sources. 2015;286:1–9.

138. Huang Y, Li H, Wang Z, Zhu M, Pei Z, Xue Q, et al. Nanostructured polypyrrole as a flexible electrode material of supercapacitor. Nano Energy. 2016;22:422–38.

139. Luo T, Xu X, Jiang M, Lu Y-z, Meng H, Li C-x. Polyacetylene carbon materials: facile preparation using AlCl 3 catalyst and excellent electrochemical performance for supercapacitors. RSC Advances. 2019;9(21):11986–95.

140. Meng Q, Cai K, Chen Y, Chen L. Research progress on conducting polymer based supercapacitor electrode materials. Nano Energy. 2017;36:268–85.

141. Muzaffar A, Ahamed MB, Deshmukh K, Thirumalai J. A review on recent advances in hybrid supercapacitors: design, fabrication and applications. Renewable and Sustainable Energy Reviews. 2019;101:123–45.

142. Afif A, Rahman SM, Azad AT, Zaini J, Islan MA, Azad AK. Advanced materials and technologies for hybrid supercapacitors for energy storage–a review. Journal of Energy Storage. 2019;25:100852.

143. Wang D-G, Liang Z, Gao S, Qu C, Zou R. Metal-organic framework-based materials for hybrid supercapacitor application. Coordination Chemistry Reviews. 2020;404:213093.

144. Najib S, Erdem E. Current progress achieved in novel materials for supercapacitor electrodes: mini review. Nanoscale Advances. 2019;1(8):2817–27.

145. Vangari M, Pryor T, Jiang L. Supercapacitors: review of materials and fabrication methods. Journal of Energy Engineering. 2013;139(2):72–79.

146. Dong L, Xu C, Li Y, Huang Z-H, Kang F, Yang Q-H, et al. Flexible electrodes and supercapacitors for wearable energy storage: a review by category. Journal of Materials Chemistry A. 2016;4(13):4659–85.

147. Yu D, Qian Q, Wei L, Jiang W, Goh K, Wei J, et al. Emergence of fiber supercapacitors. Chemical Society Reviews. 2015;44(3):647–62.

148. Fu Y, Cai X, Wu H, Lv Z, Hou S, Peng M, et al. Fiber supercapacitors utilizing pen ink for flexible/wearable energy storage. Advanced Materials. 2012;24(42):5713–18.

149. Zhai S, Jiang W, Wei L, Karahan HE, Yuan Y, Ng AK, et al. All-carbon solid-state yarn supercapacitors from activated carbon and carbon fibers for smart textiles. Materials Horizons. 2015;2(6):598–605.

150. Kim C, Choi Y-O, Lee W-J, Yang K-S. Supercapacitor performances of activated carbon fiber webs prepared by electrospinning of PMDA-ODA poly (amic acid) solutions. Electrochimica Acta. 2004; 50 (2–3): 883–87.

151. Ma W, Chen S, Yang S, Chen W, Weng W, Zhu M. Bottom-up fabrication of activated carbon fiber for all-solid-state supercapacitor with excellent electrochemical performance. ACS Applied Materials & Interfaces. 2016;8(23):14622–27.

152. Ren J, Li L, Chen C, Chen X, Cai Z, Qiu L, et al. Twisting carbon nanotube fibers for both wire-shaped micro-supercapacitor and micro-battery. Advanced Materials. 2013;25(8):1155–59.

153. Choi C, Lee JA, Choi AY, Kim YT, Lepró X, Lima MD, et al. Flexible supercapacitor made of carbon nanotube yarn with internal pores. Advanced Materials. 2014;26(13):2059–65.

154. Xu Z, Gao C. Graphene chiral liquid crystals and macroscopic assembled fibres. Nature Communications. 2011;2(1):1–9.

155. Hu Y, Cheng H, Zhao F, Chen N, Jiang L, Feng Z, et al. All-in-one graphene fiber supercapacitor. Nanoscale. 2014;6(12):6448–51.

156. Li G-X, Hou P-X, Luan J, Li J-C, Li X, Wang H, et al. A MnO2 nanosheet/single-wall carbon nanotube hybrid fiber for wearable solid-state supercapacitors. Carbon. 2018;140:634–43.
157. Chen Q, Meng Y, Hu C, Zhao Y, Shao H, Chen N, et al. MnO2-modified hierarchical graphene fiber electrochemical supercapacitor. Journal of Power Sources. 2014;247:32–39.
158. Gopalsamy K, Xu Z, Zheng B, Huang T, Kou L, Zhao X, et al. Bismuth oxide nanotubes–graphene fiber-based flexible supercapacitors. Nanoscale. 2014;6(15):8595–600.
159. Amir FZ, Pham V, Mullinax D, Dickerson J. Enhanced performance of HRGO-RuO2 solid state flexible supercapacitors fabricated by electrophoretic deposition. Carbon. 2016;107:338–43.
160. Li H, He J, Cao X, Kang L, He X, Xu H, et al. All solid-state V2O5-based flexible hybrid fiber supercapacitors. Journal of Power Sources. 2017;371:18–25.
161. Hu Y, Guan C, Ke Q, Yow ZF, Cheng C, Wang J. Hybrid Fe2O3 nanoparticle clusters/rGO paper as an effective negative electrode for flexible supercapacitors. Chemistry of Materials. 2016;28(20):7296–303.
162. Huang J, Xu Y, Xiao Y, Zhu H, Wei J, Chen Y. Mussel-inspired, biomimetics-assisted self-assembly of Co3O4 on carbon fibers for flexible supercapacitors. ChemElectroChem. 2017;4 (9):2269–77.
163. Shown I, Ganguly A, Chen LC, Chen KH. Conducting polymer-based flexible supercapacitor. Energy Science & Engineering. 2015;3(1):2–26.
164. Wang K, Meng Q, Zhang Y, Wei Z, Miao M. High-performance two-ply yarn supercapacitors based on carbon nanotubes and polyaniline nanowire arrays. Advanced Materials. 2013;25 (10):1494–98.
165. Qu G, Cheng J, Li X, Yuan D, Chen P, Chen X, et al. A fiber supercapacitor with high energy density based on hollow graphene/conducting polymer fiber electrode. Advanced Materials. 2016;28(19):3646–52.
166. Teng W, Zhou Q, Wang X, Che H, Hu P, Li H, et al. Hierarchically interconnected conducting polymer hybrid fiber with high specific capacitance for flexible fiber-shaped supercapacitor. Chemical Engineering Journal. 2020;390:124569.
167. Levitt A, Zhang J, Dion G, Gogotsi Y, Razal JM. MXene-Based fibers, yarns, and fabrics for wearable energy storage devices. Advanced Functional Materials. 2020;30(47):2000739.
168. Zhang J, Kong N, Uzun S, Levitt A, Seyedin S, Lynch PA, et al. MXene Films: scalable manufacturing of free-standing, strong Ti3C2Tx MXene films with outstanding conductivity (Adv. Mater. 23/2020). Advanced Materials. 2020;32(23):2070180.
169. Lukatskaya MR, Kota S, Lin Z, Zhao M-Q, Shpigel N, Levi MD, et al. Ultra-high-rate pseudocapacitive energy storage in two-dimensional transition metal carbides. Nature Energy. 2017;2(8):1–6.
170. Mariano M, Mashtalir O, Antonio FQ, Ryu W-H, Deng B, Xia F, et al. Solution-processed titanium carbide MXene films examined as highly transparent conductors. Nanoscale. 2016;8 (36):16371–78.
171. Yang W, Yang J, Byun JJ, Moissinac FP, Xu J, Haigh SJ, et al. 3D printing of freestanding MXene architectures for current-collector-free supercapacitors. Advanced Materials. 2019;31 (37):1902725.
172. Weng G-M, Mariano M, Lipton J, Taylor AD. MXene Films, Coatings, and Bulk Processing. 2D Metal Carbides and Nitrides (MXenes): Springer; 2019. 197–219.
173. Yang Q, Xu Z, Fang B, Huang T, Cai S, Chen H, et al. MXene/graphene hybrid fibers for high performance flexible supercapacitors. Journal of Materials Chemistry A. 2017;5(42):22113–19.
174. Liu Q, Yang J, Luo X, Miao Y, Zhang Y, Xu W, et al. Fabrication of a fibrous MnO2@ MXene/CNT electrode for high-performance flexible supercapacitor. Ceramics International. 2020;46 (8):11874–81.

175. Levitt AS, Alhabeb M, Hatter CB, Sarycheva A, Dion G, Gogotsi Y. Electrospun MXene/carbon nanofibers as supercapacitor electrodes. Journal of Materials Chemistry A. 2019;7(1):269–77.
176. Zhang J, Seyedin S, Qin S, Wang Z, Moradi S, Yang F, et al. Highly conductive Ti3C2Tx MXene hybrid fibers for flexible and elastic fiber-shaped supercapacitors. Small. 2019;15(8):1804732.
177. Zhou Z, Panatdasirisuk W, Mathis TS, Anasori B, Lu C, Zhang X, et al. Layer-by-layer assembly of MXene and carbon nanotubes on electrospun polymer films for flexible energy storage. Nanoscale. 2018;10(13):6005–13.
178. Seyedin S, Yanza ERS, Razal JM. Knittable energy storing fiber with high volumetric performance made from predominantly MXene nanosheets. Journal of Materials Chemistry A. 2017;5(46):24076–82.
179. Zhang J, Seyedin S, Gu Z, Yang W, Wang X, Razal JM. MXene: a potential candidate for yarn supercapacitors. Nanoscale. 2017;9(47):18604–08.
180. Yu C, Gong Y, Chen R, Zhang M, Zhou J, An J, et al. A solid-state fibriform supercapacitor boosted by host–guest hybridization between the carbon nanotube scaffold and MXene nanosheets. Small. 2018;14(29):1801203.
181. Wang Z, Qin S, Seyedin S, Zhang J, Wang J, Levitt A, et al. High-performance biscrolled MXene/carbon nanotube yarn supercapacitors. Small. 2018;14(37):1802225.
182. Hu M, Li Z, Li G, Hu T, Zhang C, Wang X. All-solid-state flexible fiber-based MXene supercapacitors. Advanced Materials Technologies. 2017;2(10):1700143.
183. Shao W, Tebyetekerwa M, Marriam I, Li W, Wu Y, Peng S, et al. Polyester@ MXene nanofibers-based yarn electrodes. Journal of Power Sources. 2018;396:683–90.
184. Uzun S, Seyedin S, Stoltzfus AL, Levitt AS, Alhabeb M, Anayee M, et al. Knittable and washable multifunctional mxene-coated cellulose yarns. Advanced Functional Materials. 2019;29(45):1905015.
185. Hu M, Hu T, Cheng R, Yang J, Cui C, Zhang C, et al. MXene-coated silk-derived carbon cloth toward flexible electrode for supercapacitor application. Journal of Energy Chemistry. 2018;27(1):161–66.
186. Levitt A, Hegh D, Phillips P, Uzun S, Anayee M, Razal JM, et al. 3D knitted energy storage textiles using MXene-coated yarns. Materials Today. 2020;34:17–29.
187. Jiang Q, Kurra N, Alhabeb M, Gogotsi Y, Alshareef HN. All pseudocapacitive MXene-RuO2 asymmetric supercapacitors. Advanced Energy Materials. 2018;8(13):1703043.
188. Yang Z, Deng J, Chen X, Ren J, Peng H. A highly stretchable, fiber-shaped supercapacitor. Angewandte Chemie. 2013;125(50):13695–99.
189. Yu J, Zhou J, Yao P, Huang J, Sun W, Zhu C, et al. A stretchable high performance all-in-one fiber supercapacitor. Journal of Power Sources. 2019;440:227150.
190. Sun J, Huang Y, Fu C, Wang Z, Huang Y, Zhu M, et al. High-performance stretchable yarn supercapacitor based on PPy@ CNTs@ urethane elastic fiber core spun yarn. Nano Energy. 2016;27:230–37.
191. Rajendran V, Mohan AV, Jayaraman M, Nakagawa T. All-printed, interdigitated, freestanding serpentine interconnects based flexible solid state supercapacitor for self powered wearable electronics. Nano Energy. 2019;65:104055.
192. Yu J, Lu W, Smith JP, Booksh KS, Meng L, Huang Y, et al. A high performance stretchable asymmetric fiber-shaped supercapacitor with a core-sheath helical structure. Advanced Energy Materials. 2017;7(3):1600976.
193. Noh J, Yoon C-M, Kim YK, Jang J. High performance asymmetric supercapacitor twisted from carbon fiber/MnO2 and carbon fiber/MoO3. Carbon. 2017;116:470–78.
194. Liu W, Liu N, Shi Y, Chen Y, Yang C, Tao J, et al. A wire-shaped flexible asymmetric supercapacitor based on carbon fiber coated with a metal oxide and a polymer. Journal of Materials Chemistry A. 2015;3(25):13461–67.

195. Gong X, Li S, Lee PS. A fiber asymmetric supercapacitor based on FeOOH/PPy on carbon fibers as an anode electrode with high volumetric energy density for wearable applications. Nanoscale. 2017;9(30):10794–801.
196. Cheng X, Zhang J, Ren J, Liu N, Chen P, Zhang Y, et al. Design of a hierarchical ternary hybrid for a fiber-shaped asymmetric supercapacitor with high volumetric energy density. The Journal of Physical Chemistry C. 2016;120(18):9685–91.
197. Mackanic DG, Chang T-H, Huang Z, Cui Y, Bao Z. Stretchable electrochemical energy storage devices. Chemical Society Reviews. 2020;49(13):4466–95.
198. Yu M, Zhang Y, Zeng Y, Balogun MS, Mai K, Zhang Z, et al. Water surface assisted synthesis of large-scale carbon nanotube film for high-performance and stretchable supercapacitors. Advanced Materials. 2014;26(27):4724–29.
199. Medeiros M, Helene P, Selmo S. Influence of EVA and acrylate polymers on some mechanical properties of cementitious repair mortars. Construction and Building Materials. 2009;23 (7):2527–33.
200. Zhang Z, Zhang Y, Yang K, Yi K, Zhou Z, Huang A, et al. Three-dimensional carbon nanotube/ ethylvinylacetate/polyaniline as a high performance electrode for supercapacitors. Journal of Materials Chemistry A. 2015;3(5):1884–89.
201. You I, Kong M, Jeong U. Block copolymer elastomers for stretchable electronics. Accounts of Chemical Research. 2018;52(1):63–72.
202. Liu T, Liu G. Block copolymers for supercapacitors, dielectric capacitors and batteries. Journal of Physics: Condensed Matter. 2019;31(23):233001.
203. Dong W, Wang Z, Zhang Q, Ravi M, Yu M, Tan Y, et al. Polymer/block copolymer blending system as the compatible precursor system for fabrication of mesoporous carbon nanofibers for supercapacitors. Journal of Power Sources. 2019;419:137–47.
204. Kim TH, Choi WM, Kim DH, Meitl MA, Menard E, Jiang H, et al. Printable, flexible, and stretchable forms of ultrananocrystalline diamond with applications in thermal management. Advanced Materials. 2008;20(11):2171–76.
205. Shankar R, Ghosh TK, Spontak RJ. Electromechanical response of nanostructured polymer systems with no mechanical pre-strain. Macromolecular Rapid Communications. 2007;28 (10):1142–47.
206. Yu C, Masarapu C, Rong J, Wei B, Jiang H. Stretchable supercapacitors based on buckled single-walled carbon-nanotube macrofilms. Advanced Materials. 2009;21(47):4793–97.
207. Niu Z, Dong H, Zhu B, Li J, Hng HH, Zhou W, et al. Highly stretchable, integrated supercapacitors based on single-walled carbon nanotube films with continuous reticulate architecture. Advanced Materials. 2013;25(7):1058–64.
208. Yu J, Lu W, Pei S, Gong K, Wang L, Meng L, et al. Omnidirectionally stretchable high-performance supercapacitor based on isotropic buckled carbon nanotube films. ACS Nano. 2016;10(5):5204–11.
209. Kim D, Shin G, Kang YJ, Kim W, Ha JS. Fabrication of a stretchable solid-state micro-supercapacitor array. ACS Nano. 2013;7(9):7975–82.
210. Tung K, Wong W, Pun E. Polymeric optical waveguides using direct ultraviolet photolithography process. Applied Physics A. 2005;80(3):621–26.
211. Lee C-Y, Taylor AC, Nattestad A, Beirne S, Wallace GG. 3D printing for electrocatalytic applications. Joule. 2019;3(8):1835–49.
212. Li L, Lou Z, Han W, Chen D, Jiang K, Shen G. Highly stretchable micro-supercapacitor arrays with hybrid MWCNT/PANI electrodes. Advanced Materials Technologies. 2017;2(3):1600282.
213. Shin D, Shen C, Sanghadasa M, Lin L. Breathable 3D supercapacitors based on activated carbon fiber veil. Advanced Materials Technologies. 2018;3(11):1800209.

214. Sun P, Qiu M, Li M, Mai W, Cui G, Tong Y. Stretchable Ni@ NiCoP textile for wearable energy storage clothes. Nano Energy. 2019;55:506–15.
215. Yi F, Wang J, Wang X, Niu S, Li S, Liao Q, et al. Stretchable and waterproof self-charging power system for harvesting energy from diverse deformation and powering wearable electronics. ACS Nano. 2016;10(7):6519–25.
216. Jost K, Stenger D, Perez CR, McDonough JK, Lian K, Gogotsi Y, et al. Knitted and screen printed carbon-fiber supercapacitors for applications in wearable electronics. Energy & Environmental Science. 2013;6(9):2698–705.
217. Zhao Y, Dong D, Wang Y, Gong S, An T, Yap LW, et al. Highly stretchable fiber-shaped supercapacitors based on ultrathin gold nanowires with double-helix winding design. ACS Applied Materials & Interfaces. 2018;10(49):42612–20.
218. Choi C, Lee JM, Kim SH, Kim SJ, Di J, Baughman RH. Twistable and stretchable sandwich structured fiber for wearable sensors and supercapacitors. Nano Letters. 2016;16(12):7677–84.
219. Zhang N, Zhou W, Zhang Q, Luan P, Cai L, Yang F, et al. Biaxially stretchable supercapacitors based on the buckled hybrid fiber electrode array. Nanoscale. 2015;7(29):12492–97.
220. Jin H, Zhou L, Mak CL, Huang H, Tang WM, Chan HLW. High-performance fiber-shaped supercapacitors using carbon fiber thread (CFT)@ polyanilne and functionalized CFT electrodes for wearable/stretchable electronics. Nano Energy. 2015;11:662–70.
221. Pu J, Wang X, Xu R, Komvopoulos K. Highly stretchable microsupercapacitor arrays with honeycomb structures for integrated wearable electronic systems. ACS Nano. 2016;10 (10):9306–15.
222. He S, Cao J, Xie S, Deng J, Gao Q, Qiu L, et al. Stretchable supercapacitor based on a cellular structure. Journal of Materials Chemistry A. 2016;4(26):10124–29.
223. Dong K, Wang Y-C, Deng J, Dai Y, Zhang SL, Zou H, et al. A highly stretchable and washable all-yarn-based self-charging knitting power textile composed of fiber triboelectric nanogenerators and supercapacitors. ACS Nano. 2017;11(9):9490–99.
224. Kim D, Kim D, Lee H, Jeong YR, Lee SJ, Yang G, et al. Body-attachable and stretchable multisensors integrated with wirelessly rechargeable energy storage devices. Advanced Materials. 2016;28(4):748–56.
225. Yun TG, Park M, Kim D-H, Kim D, Cheong JY, Bae JG, et al. All-transparent stretchable electrochromic supercapacitor wearable patch device. ACS Nano. 2019;13(3):3141–50.
226. Yun J, Song C, Lee H, Park H, Jeong YR, Kim JW, et al. Stretchable array of high-performance micro-supercapacitors charged with solar cells for wireless powering of an integrated strain sensor. Nano Energy. 2018;49:644–54.
227. Song W, Zhu J, Gan B, Zhao S, Wang H, Li C, et al. Flexible, stretchable, and transparent planar microsupercapacitors based on 3D porous laser-induced graphene. Small. 2018;14 (1):1702249.
228. Jiang Q, Wu C, Wang Z, Wang AC, He J-H, Wang ZL, et al. MXene electrochemical microsupercapacitor integrated with triboelectric nanogenerator as a wearable self-charging power unit. Nano Energy. 2018;45:266–72.
229. Guo H, Yeh M-H, Lai Y-C, Zi Y, Wu C, Wen Z, et al. All-in-one shape-adaptive self-charging power package for wearable electronics. ACS Nano. 2016;10(11):10580–88.
230. Kim H, Yoon J, Lee G, Paik S-h, Choi G, Kim D, et al. Encapsulated, high-performance, stretchable array of stacked planar micro-supercapacitors as waterproof wearable energy storage devices. ACS Applied Materials & Interfaces. 2016;8(25):16016–25.
231. Lee G, Kim D, Kim D, Oh S, Yun J, Kim J, et al. Fabrication of a stretchable and patchable array of high performance micro-supercapacitors using a non-aqueous solvent based gel electrolyte. Energy & Environmental Science. 2015;8(6):1764–74.

232. Hong S, Lee J, Do K, Lee M, Kim JH, Lee S, et al. Stretchable electrode based on laterally combed carbon nanotubes for wearable energy harvesting and storage devices. Advanced Functional Materials. 2017;27(48):1704353.

233. An T, Ling Y, Gong S, Zhu B, Zhao Y, Dong D, et al. A Wearable Second Skin-Like Multifunctional Supercapacitor with Vertical Gold Nanowires and Electrochromic Polyaniline. Advanced Materials Technologies. 2019;4(3):1800473.

234. Liu Z, Wu ZS, Yang S, Dong R, Feng X, Müllen K. Ultraflexible in-plane micro-supercapacitors by direct printing of solution-processable electrochemically exfoliated graphene. Advanced Materials. 2016;28(11):2217–22.

235. Luan P, Zhang N, Zhou W, Niu Z, Zhang Q, Cai L, et al. Epidermal supercapacitor with high performance. Advanced Functional Materials. 2016;26(45):8178–84.

236. Gall OZ, Meng C, Bhamra H, Mei H, John SW, Irazoqui PP. A batteryless energy harvesting storage system for implantable medical devices demonstrated in situ. Circuits, Systems, and Signal Processing. 2019;38(3):1360–73.

237. He S, Hu Y, Wan J, Gao Q, Wang Y, Xie S, et al. Biocompatible carbon nanotube fibers for implantable supercapacitors. Carbon. 2017;122:162–67.

238. Sim HJ, Choi C, Lee DY, Kim H, Yun J-H, Kim JM, et al. Biomolecule based fiber supercapacitor for implantable device. Nano Energy. 2018;47:385–92.

239. Vincent C, Scrosati B. Modern Batteries: Elsevier; 1997.

240. Linden D. Handbook of Batteries and Fuel Cells. New York. 1984.

241. Dell R, Rand DAJ. Understanding batteries: Royal society of chemistry; 2001.

242. Long JW, Dunn B, Rolison DR, White HS. Three-dimensional battery architectures. Chemical Reviews. 2004;104(10):4463–92.

243. Feng X. Analysis and simulation of a state-space based battery energy storage system. 2015.

244. Zhang Y, Zhao Y, Ren J, Weng W, Peng H. Advances in wearable fiber-shaped lithium-ion batteries. Advanced Materials. 2016;28(22):4524–31.

245. Sun Z, Jin S, Jin H, Du Z, Zhu Y, Cao A, et al. Robust expandable carbon nanotube scaffold for ultrahigh-capacity lithium-metal anodes. Advanced Materials. 2018;30(32):1800884.

246. Barré A, Deguilhem B, Grolleau S, Gérard M, Suard F, Riu D. A review on lithium-ion battery ageing mechanisms and estimations for automotive applications. Journal of Power Sources. 2013;241:680–89.

247. Park J, Jeong J, Lee Y, Oh M, Ryou MH, Lee YM. Micro-patterned lithium metal anodes with suppressed dendrite formation for post lithium-ion batteries. Advanced Materials Interfaces. 2016;3(11):1600140.

248. Selis LA, Seminario JM. Dendrite formation in silicon anodes of lithium-ion batteries. RSC Advances. 2018;8(10):5255–67.

249. Kim S-H, Choi K-H, Cho S-J, Kil E-H, Lee S-Y. Mechanically compliant and lithium dendrite growth-suppressing composite polymer electrolytes for flexible lithium-ion batteries. Journal of Materials Chemistry A. 2013;1(16):4949–55.

250. Girishkumar G, McCloskey B, Luntz AC, Swanson S, Wilcke W. Lithium– air battery: promise and challenges. The Journal of Physical Chemistry Letters. 2010;1(14):2193–203.

251. Shen X, Liu H, Cheng X-B, Yan C, Huang J-Q. Beyond lithium ion batteries: higher energy density battery systems based on lithium metal anodes. Energy Storage Materials. 2018;12:161–75.

252. Luntz A. Beyond Lithium Ion Batteries. ACS Publications; 2015.

253. Thackeray MM, Wolverton C, Isaacs ED. Electrical energy storage for transportation – approaching the limits of, and going beyond, lithium-ion batteries. Energy & Environmental Science. 2012;5(7):7854–63.

254. Hassoun J, Scrosati B. Advances in anode and electrolyte materials for the progress of lithium-ion and beyond lithium-ion batteries. Journal of The Electrochemical Society. 2015;162(14):A2582.

255. Biemolt J, Jungbacker P, Van Teijlingen T, Yan N, Rothenberg G. Beyond Lithium-based batteries. Materials. 2020;13(2):425.

256. Zhou Y, Wang CH, Lu W, Dai L. Recent advances in fiber-shaped supercapacitors and lithium-ion batteries. Advanced Materials. 2020;32(5):1902779.

257. Mo F, Liang G, Huang Z, Li H, Wang D, Zhi C. An overview of fiber-shaped batteries with a focus on multifunctionality, scalability, and technical difficulties. Advanced Materials. 2020;32(5):1902151.

258. Yang S, Cheng Y, Xiao X, Pang H. Development and application of carbon fiber in batteries. Chemical Engineering Journal. 2020;384:123294.

259. Liu B, Zhang J-G, Xu W. Advancing lithium metal batteries. Joule. 2018;2(5):833–45.

260. Wu Y, Jiang C, Wan C, Holze R. Modified natural graphite as anode material for lithium ion batteries. Journal of Power Sources. 2002;111(2):329–34.

261. Zhang X, Qu H, Ji W, Zheng D, Ding T, Abegglen C, et al. Fast and controllable prelithiation of hard carbon anodes for lithium-ion batteries. ACS Applied Materials & Interfaces. 2020;12 (10):11589–99.

262. Sandhya C, John B, Gouri C. Lithium titanate as anode material for lithium-ion cells: a review. Ionics. 2014;20(5):601–20.

263. Liu Z, Yu Q, Zhao Y, He R, Xu M, Feng S, et al. Silicon oxides: a promising family of anode materials for lithium-ion batteries. Chemical Society Reviews. 2019;48(1):285–309.

264. Feng K, Li M, Liu W, Kashkooli AG, Xiao X, Cai M, et al. Silicon-based anodes for lithium-ion batteries: from fundamentals to practical applications. Small. 2018;14(8):1702737.

265. Kamali AR, Fray DJ. Tin-based materials as advanced anode materials for lithium ion batteries: a review. Reviews on Advanced Material Science. 2011;27(1):14–24.

266. Wang X, Wang X, Lu Y. Realizing high voltage lithium cobalt oxide in lithium-ion batteries. Industrial & Engineering Chemistry Research. 2019;58(24):10119–39.

267. Berhe GB, Su W-N, Huang C-J, Hagos TM, Hagos TT, Bezabh HK, et al. A new class of lithium-ion battery using sulfurized carbon anode from polyacrylonitrile and lithium manganese oxide cathode. Journal of Power Sources. 2019;434:126641.

268. Liu S, Xiong L, He C. Long cycle life lithium ion battery with lithium nickel cobalt manganese oxide (NCM) cathode. Journal of Power Sources. 2014;261:285–91.

269. Hassoun J, Bonaccorso F, Agostini M, Angelucci M, Betti MG, Cingolani R, et al. An advanced lithium-ion battery based on a graphene anode and a lithium iron phosphate cathode. Nano Letters. 2014;14(8):4901–06.

270. Pan G, Cao F, Xia X, Zhang Y. Exploring hierarchical FeS2/C composite nanotubes arrays as advanced cathode for lithium ion batteries. Journal of Power Sources. 2016;332:383–88.

271. Zhang X-F, Wang K-X, Wei X, Chen J-S. Carbon-coated V2O5 nanocrystals as high performance cathode material for lithium ion batteries. Chemistry of Materials. 2011;23(24):5290–92.

272. Botte GG, White RE, Zhang Z. Thermal stability of LiPF6-EC: EMC electrolyte for lithium ion batteries. Journal of Power Sources. 2001;97:570–75.

273. Marom R, Haik O, Aurbach D, Halalay IC. Revisiting LiClO4 as an electrolyte for rechargeable lithium-ion batteries. Journal of the Electrochemical Society. 2010;157(8):A972.

274. Blomgren GE. The development and future of lithium ion batteries. Journal of The Electrochemical Society. 2016;164(1):A5019.

275. Xue S, Liu Y, Li Y, Teeters D, Crunkleton DW, Wang S. Diffusion of lithium ions in amorphous and crystalline poly (ethylene oxide) 3: LiCF3SO3 polymer electrolytes. Electrochimica Acta. 2017;235:122–28.

276. Mishra A, Mehta A, Basu S, Malode SJ, Shetti NP, Shukla SS, et al. Electrode materials for lithium-ion batteries. Materials Science for Energy Technologies. 2018;1(2):182–87.

277. Owens B, Reale P, Scrosati B. PRIMARY BATTERIES| Overview. 2009.

278. Yu Y, Che H, Yang X, Deng Y, Li L, Ma Z-F. Non-flammable organic electrolyte for sodium-ion batteries. Electrochemistry Communications. 2020;110:106635.

279. Lin H, Weng W, Ren J, Qiu L, Zhang Z, Chen P, et al. Twisted aligned carbon nanotube/silicon composite fiber anode for flexible wire-shaped lithium-ion battery. Advanced Materials. 2014;26(8):1217–22.

280. Zhang S, Koziol KK, Kinloch IA, Windle AH. Macroscopic fibers of well-aligned carbon nanotubes by wet spinning. Small. 2008;4(8):1217–22.

281. Lu W, Zu M, Byun JH, Kim BS, Chou TW. State of the art of carbon nanotube fibers: opportunities and challenges. Advanced materials. 2012;24(14):1805–33.

282. Li Y-L, Kinloch IA, Windle AH. Direct spinning of carbon nanotube fibers from chemical vapor deposition synthesis. Science. 2004;304(5668):276–78.

283. Koziol K, Vilatela J, Moisala A, Motta M, Cunniff P, Sennett M, et al. High-performance carbon nanotube fiber. Science. 2007;318(5858):1892–95.

284. Zhang T, Han S, Guo W, Hou F, Liu J, Yan X, et al. Continuous carbon nanotube composite fibers for flexible aqueous lithium-ion batteries. Sustainable Materials and Technologies. 2019;20:e00096.

285. Chen S, Gordin ML, Yi R, Howlett G, Sohn H, Wang D. Silicon core–hollow carbon shell nanocomposites with tunable buffer voids for high capacity anodes of lithium-ion batteries. Physical Chemistry Chemical Physics. 2012;14(37):12741–45.

286. Wang B, Li X, Zhang X, Luo B, Zhang Y, Zhi L. Contact-engineered and void-involved silicon/carbon nanohybrids as lithium-ion-battery anodes. Advanced Materials. 2013;25(26):3560–65.

287. Guo J, Chen X, Wang C. Carbon scaffold structured silicon anodes for lithium-ion batteries. Journal of Materials Chemistry. 2010;20(24):5035–40.

288. Wang C, Chui Y-S, Ma R, Wong T, Ren J-G, Wu Q-H, et al. A three-dimensional graphene scaffold supported thin film silicon anode for lithium-ion batteries. Journal of Materials Chemistry A. 2013;1(35):10092–98.

289. Weng W, Sun Q, Zhang Y, Lin H, Ren J, Lu X, et al. Winding aligned carbon nanotube composite yarns into coaxial fiber full batteries with high performances. Nano Letters. 2014;14(6):3432–38.

290. Hu L, La Mantia F, Wu H, Xie X, McDonough J, Pasta M, et al. Lithium-ion textile batteries with large areal mass loading. Advanced Energy Materials. 2011;1(6):1012–17.

291. Chong WG, Huang JQ, Xu ZL, Qin X, Wang X, Kim JK. Lithium–sulfur battery cable made from ultralight, flexible graphene/carbon nanotube/sulfur composite fibers. Advanced Functional Materials. 2017;27(4):1604815.

292. Sun C-F, Zhu H, Baker III EB, Okada M, Wan J, Ghemes A, et al. Weavable high-capacity electrodes. Nano Energy. 2013;2(5):987–94.

293. Gwon H, Hong J, Kim H, Seo D-H, Jeon S, Kang K. Recent progress on flexible lithium rechargeable batteries. Energy & Environmental Science. 2014;7(2):538–51.

294. Zhou G, Li F, Cheng H-M. Progress in flexible lithium batteries and future prospects. Energy & Environmental Science. 2014;7(4):1307–38.

295. Pan S, Yang Z, Chen P, Fang X, Guan G, Zhang Z, et al. Carbon nanostructured fibers as counter electrodes in wire-shaped dye-sensitized solar cells. The Journal of Physical Chemistry C. 2014;118(30):16419–25.

296. Pan S, Ren J, Fang X, Peng H. Integration: an effective strategy to develop multifunctional energy storage devices. Advanced Energy Materials. 2016;6(4):1501867.

297. Pu X, Song W, Liu M, Sun C, Du C, Jiang C, et al. Wearable power-textiles by integrating fabric triboelectric nanogenerators and fiber-shaped dye-sensitized solar cells. Advanced Energy Materials. 2016;6(20):1601048.

298. Tran K, Nguyen T, Bartrom A, Sadiki A, Haan J. A fuel-flexible alkaline direct liquid fuel cell. Fuel Cells. 2014;14(6):834–41.

299. Ye L, Hong Y, Liao M, Wang B, Wei D, Peng H, et al. Recent advances in flexible fiber-shaped metal-air batteries. Energy Storage Materials. 2020;28:364–74.

300. Cheng F, Chen J. Metal–air batteries: from oxygen reduction electrochemistry to cathode catalysts. Chemical Society Reviews. 2012;41(6):2172–92.

301. Wang ZL, Xu D, Xu JJ, Zhang LL, Zhang XB. Graphene oxide gel-derived, free-standing, hierarchically porous carbon for high-capacity and high-rate rechargeable Li-O2 batteries. Advanced Functional Materials. 2012;22(17):3699–705.

302. Ma S, Wu Y, Wang J, Zhang Y, Zhang Y, Yan X, et al. Reversibility of noble metal-catalyzed aprotic Li-O2 batteries. Nano letters. 2015;15(12):8084–90.

303. Oh D, Virwani K, Tadesse L, Jurich M, Aetukuri N, Thompson LE, et al. Effect of transition metal oxide cathodes on the oxygen evolution reaction in Li–O2 batteries. The Journal of Physical Chemistry C. 2017;121(3):1404–11.

304. Lv Y, Li Z, Yu Y, Yin J, Song K, Yang B, et al. Copper/cobalt-doped LaMnO3 perovskite oxide as a bifunctional catalyst for rechargeable Li-O2 batteries. Journal of Alloys and Compounds. 2019;801:19–26.

305. Park J, Jun Y-S, Lee W-r, Gerbec JA, See KA, Stucky GD. Bimodal mesoporous titanium nitride/carbon microfibers as efficient and stable electrocatalysts for Li–O2 batteries. Chemistry of Materials. 2013;25(19):3779–81.

306. Xiong Q, Huang G, Zhang XB. High-capacity and stable Li-O2 batteries enabled by a trifunctional soluble redox mediator. Angewandte Chemie. 2020;132(43):19473–81.

307. Kwak WJ, Park SJ, Jung HG, Sun YK. Optimized concentration of redox mediator and surface protection of Li metal for maintenance of high energy efficiency in Li–O2 batteries. Advanced Energy Materials. 2018;8(9):1702258.

308. Shui J, Du F, Xue C, Li Q, Dai L. Vertically aligned N-doped coral-like carbon fiber arrays as efficient air electrodes for high-performance nonaqueous Li–O2 batteries. ACS Nano. 2014;8 (3):3015–22.

309. Nie Y, Li L, Wei Z. Recent advancements in Pt and Pt-free catalysts for oxygen reduction reaction. Chemical Society Reviews. 2015;44(8):2168–201.

310. Strickler AL, Jackson A, Jaramillo TF. Active and stable Ir@ Pt core–shell catalysts for electrochemical oxygen reduction. ACS Energy Letters. 2017;2(1):244–49.

311. Xiao M, Gao L, Wang Y, Wang X, Zhu J, Jin Z, et al. Engineering energy level of metal center: Ru single-atom site for efficient and durable oxygen reduction catalysis. Journal of the American Chemical Society. 2019;141(50):19800–06.

312. Wang Y, Li J, Wei Z. Transition-metal-oxide-based catalysts for the oxygen reduction reaction. Journal of Materials Chemistry A. 2018;6(18):8194–209.

313. Liu G, Li X, Lee J-W, Popov BN. A review of the development of nitrogen-modified carbon-based catalysts for oxygen reduction at USC. Catalysis Science & Technology. 2011;1(2):207–17.

314. Meng F, Zhong H, Bao D, Yan J, Zhang X. In situ coupling of strung Co4N and intertwined N–C fibers toward free-standing bifunctional cathode for robust, efficient, and flexible Zn–air batteries. Journal of the American Chemical Society. 2016;138(32):10226–31.

315. Macdonald DD, English C. Development of anodes for aluminium/air batteries – solution phase inhibition of corrosion. Journal of Applied Electrochemistry. 1990;20(3):405–17.

316. Katsoufis P, Mylona V, Politis C, Avgouropoulos G, Lianos P. Study of some basic operation conditions of an Al-air battery using technical grade commercial aluminum. Journal of Power Sources. 2020;450:227624.

317. Xu Y, Zhao Y, Ren J, Zhang Y, Peng H. An all-solid-state fiber-shaped aluminum–air battery with flexibility, stretchability, and high electrochemical performance. Angewandte Chemie. 2016;128(28):8111–14.

318. Ma Y, Sumboja A, Zang W, Yin S, Wang S, Pennycook SJ, et al. Flexible and wearable all-solid-state Al–Air battery based on iron carbide encapsulated in electrospun porous carbon nanofibers. ACS Applied Materials & Interfaces. 2018;11(2):1988–95.

319. Takechi K, Shiga T, Asaoka T. A li–o 2/co 2 battery. Chemical Communications. 2011;47 (12):3463–65.

320. Li Y, Zhou J, Zhang T, Wang T, Li X, Jia Y, et al. Highly surface-wrinkled and N-doped CNTs anchored on metal wire: a novel fiber-shaped cathode toward high-performance flexible Li–CO2 batteries. Advanced Functional Materials. 2019;29(12):1808117.

321. Ji X, Nazar LF. Advances in Li–S batteries. Journal of Materials Chemistry. 2010;20 (44):9821–26.

322. Elazari R, Salitra G, Garsuch A, Panchenko A, Aurbach D. Sulfur-impregnated activated carbon fiber cloth as a binder-free cathode for rechargeable Li-S batteries. Advanced Materials. 2011;23(47):5641–44.

323. Wu F, Shi L, Mu D, Xu H, Wu B. A hierarchical carbon fiber/sulfur composite as cathode material for Li–S batteries. Carbon. 2015;86:146–55.

324. Hwang TH, Jung DS, Kim J-S, Kim BG, Choi JW. One-dimensional carbon–sulfur composite fibers for Na–S rechargeable batteries operating at room temperature. Nano Letters. 2013;13 (9):4532–38.

325. Ding J, Zhang H, Fan W, Zhong C, Hu W, Mitlin D. Review of Emerging Potassium–Sulfur Batteries. Advanced Materials. 2020;32(23):1908007.

326. Shen C, Yuan K, Tian T, Bai M, Wang J-G, Li X, et al. Flexible sub-micro carbon fiber@ CNTs as anodes for potassium-ion batteries. ACS Applied Materials & Interfaces. 2019;11(5):5015–21.

327. Muthuraj D, Ghosh A, Kumar A, Mitra S. Nitrogen and sulfur doped carbon cloth as current collector and polysulfide immobilizer for magnesium-sulfur batteries. ChemElectroChem. 2019;6:684–89.

328. Hong X, Mei J, Wen L, Tong Y, Vasileff AJ, Wang L, et al. Nonlithium metal–sulfur batteries: steps toward a leap. Advanced Materials. 2019;31(5):1802822.

329. Yu X, Manthiram A. Electrochemical energy storage with a reversible nonaqueous room-temperature aluminum–sulfur chemistry. Advanced Energy Materials. 2017;7(18):1700561.

330. Jin Z, Li P, Jin Y, Xiao D. Superficial-defect engineered nickel/iron oxide nanocrystals enable high-efficient flexible fiber battery. Energy Storage Materials. 2018;13:160–67.

331. Hatzell KB, Chen XC, Cobb CL, Dasgupta NP, Dixit MB, Marbella LE, et al. Challenges in lithium metal anodes for solid-state batteries. ACS Energy Letters. 2020;5(3):922–34.

332. Wang X, Pan Z, Yang J, Lyu Z, Zhong Y, Zhou G, et al. Stretchable fiber-shaped lithium metal anode. Energy Storage Materials. 2019;22:179–84.

333. Zhang Y, Bai W, Ren J, Weng W, Lin H, Zhang Z, et al. Super-stretchy lithium-ion battery based on carbon nanotube fiber. Journal of Materials Chemistry A. 2014;2(29):11054–59.

334. Kumar R, Shin J, Yin L, You JM, Meng YS, Wang J. All-printed, stretchable Zn-Ag2O rechargeable battery via hyperelastic binder for self-powering wearable electronics. Advanced Energy Materials. 2017;7(8):1602096.

335. Yin L, Scharf J, Ma J, Doux J-M, Redquest C, Le VL, et al. High performance printed AgO-Zn rechargeable battery for flexible electronics. Joule. 2021;5(1):228–48.

336. Zhang Y, Xu S, Fu H, Lee J, Su J, Hwang K-C, et al. Buckling in serpentine microstructures and applications in elastomer-supported ultra-stretchable electronics with high areal coverage. Soft Matter. 2013;9(33):8062–70.

337. Mohan AV, Kim N, Gu Y, Bandodkar AJ, You JM, Kumar R, et al. Merging of Thin-and Thick-Film Fabrication Technologies: toward Soft Stretchable "Island–Bridge" Devices. Advanced Materials Technologies. 2017;2(4):1600284.

338. Xu S, Zhang Y, Cho J, Lee J, Huang X, Jia L, et al. Stretchable batteries with self-similar serpentine interconnects and integrated wireless recharging systems. Nature Communications. 2013;4(1):1–8.

339. Zamarayeva AM, Ostfeld AE, Wang M, Duey JK, Deckman I, Lechêne BP, et al. Flexible and stretchable power sources for wearable electronics. Science Advances. 2017;3(6):e1602051.

340. Song Z, Ma T, Tang R, Cheng Q, Wang X, Krishnaraju D, et al. Origami lithium-ion batteries. Nature Communications. 2014;5(1):1–6.

341. Shyu TC, Damasceno PF, Dodd PM, Lamoureux A, Xu L, Shlian M, et al. A kirigami approach to engineering elasticity in nanocomposites through patterned defects. Nature Materials. 2015;14(8):785–89.

342. Zhang Y, Bai W, Cheng X, Ren J, Weng W, Chen P, et al. Flexible and stretchable lithium-ion batteries and supercapacitors based on electrically conducting carbon nanotube fiber springs. Angewandte Chemie International Edition. 2014;53(52):14564–68.

343. Ren J, Zhang Y, Bai W, Chen X, Zhang Z, Fang X, et al. Elastic and wearable wire-shaped lithium-ion battery with high electrochemical performance. Angewandte Chemie. 2014;126 (30):7998–8003.

344. Weng W, Sun Q, Zhang Y, He S, Wu Q, Deng J, et al. A gum-like lithium-ion battery based on a novel arched structure. Advanced Materials. 2015;27(8):1363–69.

345. Liu W, Chen Z, Zhou G, Sun Y, Lee HR, Liu C, et al. 3D porous sponge-inspired electrode for stretchable lithium-ion batteries. Advanced Materials. 2016;28(18):3578–83.

346. Yan C, Wang X, Cui M, Wang J, Kang W, Foo CY, et al. Stretchable silver-zinc batteries based on embedded nanowire elastic conductors. Advanced Energy Materials. 2014;4(5):1301396.

347. Li H, Liu Z, Liang G, Huang Y, Huang Y, Zhu M, et al. Waterproof and tailorable elastic rechargeable yarn zinc ion batteries by a cross-linked polyacrylamide electrolyte. ACS Nano. 2018;12(4):3140–48.

348. Gaikwad AM, Zamarayeva AM, Rousseau J, Chu H, Derin I, Steingart DA. Highly stretchable alkaline batteries based on an embedded conductive fabric. Advanced Materials. 2012;24 (37):5071–76.

349. Kaltenbrunner M, Kettlgruber G, Siket C, Schwödiauer R, Bauer S. Arrays of ultracompliant electrochemical dry gel cells for stretchable electronics. Advanced Materials. 2010;22 (18):2065–67.

350. Li H, Ding Y, Ha H, Shi Y, Peng L, Zhang X, et al. An All-Stretchable-Component Sodium-Ion Full Battery. Advanced Materials. 2017;29(23):1700898.

351. Shin M, Song WJ, Son HB, Yoo S, Kim S, Song G, et al. Highly Stretchable Separator Membrane for Deformable Energy-Storage Devices. Advanced Energy Materials. 2018;8 (23):1801025.

352. Song WJ, Park J, Kim DH, Bae S, Kwak MJ, Shin M, et al. Jabuticaba-Inspired hybrid carbon filler/polymer electrode for use in highly stretchable aqueous Li-ion batteries. Advanced Energy Materials. 2018;8(10):1702478.

353. Wang L, Zhang Y, Pan J, Peng H. Stretchable lithium-air batteries for wearable electronics. Journal of Materials Chemistry A. 2016;4(35):13419–24.

354. Ngai KS, Ramesh S, Ramesh K, Juan JC. A review of polymer electrolytes: fundamental, approaches and applications. Ionics. 2016;22(8):1259–79.

355. Stephan AM. Review on gel polymer electrolytes for lithium batteries. European Polymer Journal. 2006;42(1):21–42.
356. Pal B, Yang S, Ramesh S, Thangadurai V, Jose R. Electrolyte selection for supercapacitive devices: a critical review. Nanoscale Advances. 2019;1(10):3807–35.
357. Xu K. Nonaqueous liquid electrolytes for lithium-based rechargeable batteries. Chemical Reviews. 2004;104(10):4303–418.
358. Wang Y, Zhong WH. Development of electrolytes towards achieving safe and high-performance energy-storage devices: a Review. ChemElectroChem. 2015;2(1):22–36.
359. Agrawal R, Pandey G. Solid polymer electrolytes: materials designing and all-solid-state battery applications: an overview. Journal of Physics D: Applied Physics. 2008;41 (22):223001.
360. Zhang Q, Liu K, Ding F, Liu X. Recent advances in solid polymer electrolytes for lithium batteries. Nano Research. 2017;10(12):4139–74.
361. Zhou Q, Ma J, Dong S, Li X, Cui G. Intermolecular chemistry in solid polymer electrolytes for high-energy-density lithium batteries. Advanced Materials. 2019;31(50):1902029.
362. Rajendran S, Sivakumar M, Subadevi R. Effect of salt concentration in poly (vinyl alcohol)-based solid polymer electrolytes. Journal of Power Sources. 2003;124(1):225–30.
363. Michael M, Jacob M, Prabaharan S, Radhakrishna S. Enhanced lithium ion transport in PEO-based solid polymer electrolytes employing a novel class of plasticizers. Solid State Ionics. 1997; 98 (3–4): 167–74.
364. Song J, Wang Y, Wan CC. Review of gel-type polymer electrolytes for lithium-ion batteries. Journal of Power Sources. 1999;77(2):183–97.
365. Hofmann A, Schulz M, Hanemann T. Gel electrolytes based on ionic liquids for advanced lithium polymer batteries. Electrochimica Acta. 2013;89:823–31.
366. Osada I, De Vries H, Scrosati B, Passerini S. Ionic-liquid-based polymer electrolytes for battery applications. Angewandte Chemie International Edition. 2016;55(2):500–13.
367. Fuller J, Breda AC, Carlin RT. Ionic liquid–polymer gel electrolytes from hydrophilic and hydrophobic ionic liquids. Journal of Electroanalytical Chemistry. 1998;459(1):29–34.
368. Fuller J, Breda A, Carlin R. Ionic liquid-polymer gel electrolytes. Journal of the Electrochemical Society. 1997;144(4):L67.
369. Egashira M, Todo H, Yoshimoto N, Morita M. Lithium ion conduction in ionic liquid-based gel polymer electrolyte. Journal of Power Sources. 2008;178(2):729–35.
370. Reiter J, Vondrák J, Michálek J, Mička Z. Ternary polymer electrolytes with 1-methylimidazole based ionic liquids and aprotic solvents. Electrochimica Acta. 2006;52(3):1398–408.
371. Fernicola A, Weise F, Greenbaum S, Kagimoto J, Scrosati B, Soleto A. Lithium-ion-conducting electrolytes: from an ionic liquid to the polymer membrane. Journal of the Electrochemical Society. 2009;156(7):A514.
372. Ye H, Huang J, Xu JJ, Khalfan A, Greenbaum SG. Li ion conducting polymer gel electrolytes based on ionic liquid/PVDF-HFP blends. Journal of the Electrochemical Society. 2007;154(11): A1048.
373. Singh B, Sekhon S. Polymer electrolytes based on room temperature ionic liquid: 2, 3-dimethyl-1-octylimidazolium triflate. The Journal of Physical Chemistry B. 2005;109 (34):16539–43.
374. Cui W-Y, An M-Z, Yang P-X. Preparation of an ionic liquid gel polymer electrolyte and its compatibility with a LiFePO4 cathode. Acta Physico-Chimica Sinica. 2010;26(5):1233–38.
375. Bansal D, Cassel F, Croce F, Hendrickson M, Plichta E, Salomon M. Conductivities and transport properties of gelled electrolytes with and without an ionic liquid for Li and Li-ion batteries. The Journal of Physical Chemistry B. 2005;109(10):4492–96.

376. Shin J-H, Henderson WA, Passerini S. PEO-based polymer electrolytes with ionic liquids and their use in lithium metal-polymer electrolyte batteries. Journal of the Electrochemical Society. 2005;152(5):A978.

377. Choi J-W, Cheruvally G, Kim Y-H, Kim J-K, Manuel J, Raghavan P, et al. Poly (ethylene oxide)-based polymer electrolyte incorporating room-temperature ionic liquid for lithium batteries. Solid State Ionics. 2007; 178 (19–20): 1235–41.

378. Song WJ, Yoo S, Song G, Lee S, Kong M, Rim J, et al. Recent progress in stretchable batteries for wearable electronics. Batteries & Supercaps. 2019;2(3):181–99.

379. Lee SS, Choi KH, Kim SH, Lee SY. Wearable supercapacitors printed on garments. Advanced Functional Materials. 2018;28(11):1705571.

380. Tamilarasan P, Ramaprabhu S. Stretchable supercapacitors based on highly stretchable ionic liquid incorporated polymer electrolyte. Materials Chemistry and Physics. 2014; 148 (1–2): 48–56.

381. Lee J, Kim W, Kim W. Stretchable carbon nanotube/ion–gel supercapacitors with high durability realized through interfacial microroughness. ACS Applied Materials & Interfaces. 2014;6(16):13578–86.

382. Peng H, Lv Y, Wei G, Zhou J, Gao X, Sun K, et al. A flexible and self-healing hydrogel electrolyte for smart supercapacitor. Journal of Power Sources. 2019;431:210–19.

383. Zhao J, Gong J, Wang G, Zhu K, Ye K, Yan J, et al. A self-healing hydrogel electrolyte for flexible solid-state supercapacitors. Chemical Engineering Journal. 2020;401:125456.

384. Lin Y, Zhang H, Liao H, Zhao Y, Li K. A physically crosslinked, self-healing hydrogel electrolyte for nano-wire PANI flexible supercapacitors. Chemical Engineering Journal. 2019;367:139–48.

385. Ye T, Li L, Zhang Y. Recent Progress in Solid Electrolytes for Energy Storage Devices. Advanced Functional Materials. 2020;30(29):2000077.

386. Hagenmuller P, Van Gool W. Solid Electrolytes: General Principles, Characterization, Materials, Applications: Elsevier; 2015.

387. Fergus JW. Ceramic and polymeric solid electrolytes for lithium-ion batteries. Journal of Power Sources. 2010;195(15):4554–69.

388. Zheng F, Kotobuki M, Song S, Lai MO, Lu L. Review on solid electrolytes for all-solid-state lithium-ion batteries. Journal of Power Sources. 2018;389:198–213.

389. Deng K, Qin J, Wang S, Ren S, Han D, Xiao M, et al. Effective Suppression of Lithium Dendrite Growth Using a Flexible Single-Ion Conducting Polymer Electrolyte. Small. 2018;14(31):1801420.

390. Takada K. Progress and prospective of solid-state lithium batteries. Acta Materialia. 2013;61 (3):759–70.

391. Fenton D. Complexes of alkali metal ions with poly (ethylene oxide). POLYmer. 1973;14:589.

392. Quartarone E, Mustarelli P, Magistris A. PEO-based composite polymer electrolytes. Solid State Ionics. 1998; 110 (1–2): 1–14.

393. Appetecchi GB, Croce F, Hassoun J, Scrosati B, Salomon M, Cassel F. Hot-pressed, dry, composite, PEO-based electrolyte membranes: I. Ionic conductivity characterization. Journal of Power Sources. 2003;114(1):105–12.

394. Scrosati B, Croce F, Persi L. Impedance spectroscopy study of PEO-based nanocomposite polymer electrolytes. Journal of the Electrochemical Society. 2000;147(5):1718.

395. Bandara L, Dissanayake M, Mellander B-E. Ionic conductivity of plasticized (PEO)-LiCF3SO3 electrolytes. Electrochimica acta. 1998; 43 (10–11): 1447–51.

396. Ahn J-H, Wang G, Liu H, Dou S. Nanoparticle-dispersed PEO polymer electrolytes for Li batteries. Journal of Power Sources. 2003;119:422–26.

397. Singh M, Odusanya O, Wilmes GM, Eitouni HB, Gomez ED, Patel AJ, et al. Effect of molecular weight on the mechanical and electrical properties of block copolymer electrolytes. Macromolecules. 2007;40(13):4578–85.

398. Rodríguez J, Navarrete E, Dalchiele EA, Sánchez L, Ramos-Barrado JR, Martín F. Polyvinylpyrrolidone–LiClO4 solid polymer electrolyte and its application in transparent thin film supercapacitors. Journal of Power Sources. 2013;237:270–76.

399. Zhao Y, Wu C, Peng G, Chen X, Yao X, Bai Y, et al. A new solid polymer electrolyte incorporating Li10GeP2S12 into a polyethylene oxide matrix for all-solid-state lithium batteries. Journal of Power Sources. 2016;301:47–53.

400. Kamaya N, Homma K, Yamakawa Y, Hirayama M, Kanno R, Yonemura M, et al. A lithium superionic conductor. Nature Materials. 2011;10(9):682–86.

401. Michot T, Nishimoto A, Watanabe M. Electrochemical properties of polymer gel electrolytes based on poly (vinylidene fluoride) copolymer and homopolymer. Electrochimica Acta. 2000; 45 (8–9): 1347–60.

402. Li W, Pang Y, Liu J, Liu G, Wang Y, Xia Y. A PEO-based gel polymer electrolyte for lithium ion batteries. RSC Advances. 2017;7(38):23494–501.

403. Walkowiak M, Zalewska A, Jesionowski T, Pokora M. Stability of poly (vinylidene fluoride-co-hexafluoropropylene)-based composite gel electrolytes with functionalized silicas. Journal of Power Sources. 2007;173(2):721–28.

404. Dissanayake M, Bandara L, Bokalawala R, Jayathilaka P, Ileperuma O, Somasundaram S. A novel gel polymer electrolyte based on polyacrylonitrile (PAN) and its application in a solar cell. Materials Research Bulletin. 2002;37(5):867–74.

405. Yang H, Huang M, Wu J, Lan Z, Hao S, Lin J. The polymer gel electrolyte based on poly (methyl methacrylate) and its application in quasi-solid-state dye-sensitized solar cells. Materials Chemistry and Physics. 2008;110(1):38–42.

406. Gao H, Zhou W, Park K, Goodenough JB. A sodium-ion battery with a low-cost cross-linked gel-polymer electrolyte. Advanced Energy Materials. 2016;6(18):1600467.

407. Shin W-K, Cho J, Kannan AG, Lee Y-S, Kim D-W. Cross-linked composite gel polymer electrolyte using mesoporous methacrylate-functionalized SiO 2 nanoparticles for lithium-ion polymer batteries. Scientific Reports. 2016;6(1):1–10.

408. Chen Q, Li X, Zang X, Cao Y, He Y, Li P, et al. Effect of different gel electrolytes on graphene-based solid-state supercapacitors. RSC Advances. 2014;4(68):36253–56.

409. Alipoori S, Mazinani S, Aboutalebi SH, Sharif F. Review of PVA-based gel polymer electrolytes in flexible solid-state supercapacitors: opportunities and challenges. Journal of Energy Storage. 2020;27:101072.

410. Ramaswamy M, Malayandi T, Subramanian S, Srinivasalu J, Rangaswamy M. Magnesium ion conducting polyvinyl alcohol–polyvinyl pyrrolidone-based blend polymer electrolyte. Ionics. 2017;23(7):1771–81.

411. Thanh HTT, Le PA, Thi MD, Le Quang T, Trinh TN. Effect of gel polymer electrolyte based on polyvinyl alcohol/polyethylene oxide blend and sodium salts on the performance of solid-state supercapacitor. Bulletin of Materials Science. 2018;41(6):1–6.

412. Patel S, Patel R, Awadhia A, Chand N, Agrawal S. Role of polyvinyl alcohol in the conductivity behaviour of polyethylene glycol-based composite gel electrolytes. Pramana. 2007;69 (3):467–75.

413. Ramasamy C, Anderson M. An activated carbon supercapacitor analysis by using a gel electrolyte of sodium salt-polyethylene oxide in an organic mixture solvent. Journal of Solid State Electrochemistry. 2014;18(8):2217–23.

414. Kang YJ, Chung H, Han C-H, Kim W. All-solid-state flexible supercapacitors based on papers coated with carbon nanotubes and ionic-liquid-based gel electrolytes. Nanotechnology. 2012;23(6):065401.

415. Basumallick I, Roy P, Chatterjee A, Bhattacharya A, Chatterjee S, Ghosh S. Organic polymer gel electrolyte for Li-ion batteries. Journal of power Sources. 2006;162(2):797–99.

416. Zaghib K, Charest P, Guerfi A, Shim J, Perrier M, Striebel K. LiFePO4 safe Li-ion polymer batteries for clean environment. Journal of Power Sources. 2005; 146 (1–2): 380–85.

417. Tarascon J-M, Gozdz A, Schmutz C, Shokoohi F, Warren P. Performance of Bellcore's plastic rechargeable Li-ion batteries. Solid State Ionics. 1996;86:49–54.

418. Aravindan V, Vickraman P, Sivashanmugam A, Thirunakaran R, Gopukumar S. LiFAP-based PVdF–HFP microporous membranes by phase-inversion technique with Li/LiFePO 4 cell. Applied Physics A. 2009;97(4):811–19.

419. Wang Y, Li B, Ji J, Eyler A, Zhong WH. A Gum-Like Electrolyte: safety of a Solid, Performance of a Liquid. Advanced Energy Materials. 2013;3(12):1557–62.

420. Wang Y, Qiao X, Zhang C, Zhou X. Development of structural supercapacitors with epoxy based adhesive polymer electrolyte. Journal of Energy Storage. 2019;26:100968.

421. Bandodkar AJ, You J-M, Kim N-H, Gu Y, Kumar R, Mohan AV, et al. Soft, stretchable, high power density electronic skin-based biofuel cells for scavenging energy from human sweat. Energy & Environmental Science. 2017;10(7):1581–89.

422. Zhao C, Wang C, Yue Z, Shu K, Wallace GG. Intrinsically stretchable supercapacitors composed of polypyrrole electrodes and highly stretchable gel electrolyte. ACS Applied Materials & Interfaces. 2013;5(18):9008–14.

423. Hu K, Xie X, Szkopek T, Cerruti M. Understanding hydrothermally reduced graphene oxide hydrogels: from reaction products to hydrogel properties. Chemistry of Materials. 2016;28 (6):1756–68.

424. Sarkar SD, Uddin MM, Roy CK, Hossen MJ, Sujan MI, Azam MS. Mechanically tough and highly stretchable poly (acrylic acid) hydrogel cross-linked by 2D graphene oxide. RSC Advances. 2020;10(18):10949–58.

425. Guan J, Li Y, Li J. Stretchable ionic-liquid-based gel polymer electrolytes for lithium-ion batteries. Industrial & Engineering Chemistry Research. 2017;56(44):12456–63.

426. Trask R, Williams H, Bond I. Self-healing polymer composites: mimicking nature to enhance performance. Bioinspiration & Biomimetics. 2007;2(1):P1.

427. Guo Y, Zhou X, Tang Q, Bao H, Wang G, Saha P. A self-healable and easily recyclable supramolecular hydrogel electrolyte for flexible supercapacitors. Journal of Materials Chemistry A. 2016;4(22):8769–76.

428. Pu X, Liu M, Li L, Zhang C, Pang Y, Jiang C, et al. Efficient charging of Li-ion batteries with pulsed output current of triboelectric nanogenerators. Advanced Science. 2016;3(1):1500255.

429. Xi F, Pang Y, Li W, Jiang T, Zhang L, Guo T, et al. Universal power management strategy for triboelectric nanogenerator. Nano Energy. 2017;37:168–76.

430. Cheng X, Pan J, Zhao Y, Liao M, Peng H. Gel polymer electrolytes for electrochemical energy storage. Advanced Energy Materials. 2018;8(7):1702184.

431. Jeon I, Cui J, Illeperuma WR, Aizenberg J, Vlassak JJ. Extremely stretchable and fast self-healing hydrogels. Advanced Materials. 2016;28(23):4678–83.

432. Li J, Geng L, Wang G, Chu H, Wei H. Self-healable gels for use in wearable devices. Chemistry of Materials. 2017;29(21):8932–52.

433. Bandodkar AJ, Mohan V, López CS, Ramírez J, Wang J. Self-healing inks for autonomous repair of printable electrochemical devices. Advanced Electronic Materials. 2015;1(12):1500289.

434. Liu S, Rao Z, Wu R, Sun Z, Yuan Z, Bai L, et al. Fabrication of microcapsules by the combination of biomass porous carbon and polydopamine for dual self-healing hydrogels. Journal of Agricultural and Food Chemistry. 2019;67(4):1061–71.

435. Zhao Y, Fickert J, Landfester K, Crespy D. Encapsulation of self-healing agents in polymer nanocapsules. Small. 2012;8(19):2954–58.

436. Cromwell OR, Chung J, Guan Z. Malleable and self-healing covalent polymer networks through tunable dynamic boronic ester bonds. Journal of the American Chemical Society. 2015;137(20):6492–95.

437. Lin Y, Li G. An intermolecular quadruple hydrogen-bonding strategy to fabricate self-healing and highly deformable polyurethane hydrogels. Journal of Materials Chemistry B. 2014;2 (39):6878–85.

438. Shen Z, Jiang Y, Wang T, Liu M. Symmetry breaking in the supramolecular gels of an achiral gelator exclusively driven by π–π stacking. Journal of the American Chemical Society. 2015;137(51):16109–15.

439. Häring M, Díaz DD. Supramolecular metallogels with bulk self-healing properties prepared by in situ metal complexation. Chemical Communications. 2016;52(89):13068–81.

440. Tuncaboylu DC, Sari M, Oppermann W, Okay O. Tough and self-healing hydrogels formed via hydrophobic interactions. Macromolecules. 2011;44(12):4997–5005.

441. Kakuta T, Takashima Y, Nakahata M, Otsubo M, Yamaguchi H, Harada A. Preorganized hydrogel: self-healing properties of supramolecular hydrogels formed by polymerization of host–guest-monomers that contain cyclodextrins and hydrophobic guest groups. Advanced Materials. 2013;25(20):2849–53.

442. Kuhl N, Bode S, Bose RK, Vitz J, Seifert A, Hoeppener S, et al. Acylhydrazones as reversible covalent crosslinkers for self-healing polymers. Advanced Functional Materials. 2015;25 (22):3295–301.

443. Han X, Meng X, Wu Z, Wu Z, Qi X. Dynamic imine bond cross-linked self-healing thermosensitive hydrogels for sustained anticancer therapy via intratumoral injection. Materials Science and Engineering: C. 2018;93:1064–72.

444. Yang WJ, Tao X, Zhao T, Weng L, Kang E-T, Wang L. Antifouling and antibacterial hydrogel coatings with self-healing properties based on a dynamic disulfide exchange reaction. Polymer Chemistry. 2015;6(39):7027–35.

445. Ye X, Li X, Shen Y, Chang G, Yang J, Gu Z. Self-healing pH-sensitive cytosine-and guanosine-modified hyaluronic acid hydrogels via hydrogen bonding. Polymer. 2017;108:348–60.

446. Zhao Y, Zhang Y, Sun H, Dong X, Cao J, Wang L, et al. A self-healing aqueous lithium-ion battery. Angewandte Chemie International Edition. 2016;55(46):14384–88.

447. Li L, Yan B, Yang J, Chen L, Zeng H. Novel mussel-inspired injectable self-healing hydrogel with anti-biofouling property. Advanced Materials. 2015;27(7):1294–99.

448. Yavvari PS, Srivastava A. Robust, self-healing hydrogels synthesised from catechol rich polymers. Journal of Materials Chemistry B. 2015;3(5):899–910.

449. Tran NB, Moon JR, Jeon YS, Kim J, Kim J-H. Adhesive and self-healing soft gel based on metal-coordinated imidazole-containing polyaspartamide. Colloid and Polymer Science. 2017;295 (4):655–64.

450. Sharma A, Rawat K, Solanki PR, Bohidar H. Self-healing gelatin ionogels. International Journal of Biological Macromolecules. 2017;95:603–07.

451. Li Y, Li J, Zhao X, Yan Q, Gao Y, Hao J, et al. Triterpenoid-based self-healing supramolecular polymer hydrogels formed by host–guest interactions. Chemistry–A European Journal. 2016;22(51):18435–41.

452. Apostolides DE, Sakai T, Patrickios CS. Dynamic covalent star poly (ethylene glycol) model hydrogels: a new platform for mechanically robust, multifunctional materials. Macromolecules. 2017;50(5):2155–64.

453. Zhao X, Zhang M, Guo B, Ma PX. Mussel-inspired injectable supramolecular and covalent bond crosslinked hydrogels with rapid self-healing and recovery properties via a facile approach under metal-free conditions. Journal of Materials Chemistry B. 2016;4(41):6644–51.

454. Pettignano A, Grijalvo S, Haering M, Eritja R, Tanchoux N, Quignard F, et al. Boronic acid-modified alginate enables direct formation of injectable, self-healing and multistimuli-responsive hydrogels. Chemical Communications. 2017;53(23):3350–53.

455. Dutta A, Maity S, Das RK. A Highly Stretchable, Tough, Self-Healing, and Thermoprocessable Polyacrylamide–Chitosan Supramolecular Hydrogel. Macromolecular Materials and Engineering. 2018;303(12):1800322.

456. Zhou B, He D, Hu J, Ye Y, Peng H, Zhou X, et al. A flexible, self-healing and highly stretchable polymer electrolyte via quadruple hydrogen bonding for lithium-ion batteries. Journal of Materials Chemistry A. 2018;6(25):11725–33.

457. Colquhoun HM, Zhu Z, Williams DJ, Drew MG, Cardin CJ, Gan Y, et al. Induced-Fit Binding of π-electron-donor substrates to macrocyclic aromatic ether imide sulfones: a versatile approach to molecular assembly. Chemistry–A European Journal. 2010;16(3):907–18.

458. Mei JF, Jia XY, Lai JC, Sun Y, Li CH, Wu JH, et al. A highly stretchable and autonomous self-healing polymer based on combination of pt⋯ pt and π–π interactions. Macromolecular Rapid Communications. 2016;37(20):1667–75.

459. Shi L, Ding P, Wang Y, Zhang Y, Ossipov D, Hilborn J. Self-healing polymeric hydrogel formed by metal–ligand coordination assembly: design, fabrication, and biomedical applications. Macromolecular Rapid Communications. 2019;40(7):1800837.

460. Hu Y, Shen P, Zeng N, Wang L, Yan D, Cui L, et al. Hybrid hydrogel electrolyte based on metal–organic supermolecular self-assembly and polymer chemical cross-linking for rechargeable aqueous Zn–MnO2 batteries. ACS Applied Materials & Interfaces. 2020;12 (37):42285–93.

461. Okay O. Self-healing hydrogels formed via hydrophobic interactions. Supramolecular Polymer Networks and Gels: Springer; 2015. 101–42.

462. Zhang M, Xu D, Yan X, Chen J, Dong S, Zheng B, et al. Self-healing supramolecular gels formed by crown ether based host–guest interactions. Angewandte Chemie. 2012;124 (28):7117–21.

463. Liu F, Li F, Deng G, Chen Y, Zhang B, Zhang J, et al. Rheological images of dynamic covalent polymer networks and mechanisms behind mechanical and self-healing properties. Macromolecules. 2012;45(3):1636–45.

464. Skene WG, Lehn J-MP. Dynamers: polyacylhydrazone reversible covalent polymers, component exchange, and constitutional diversity. Proceedings of the National Academy of Sciences. 2004;101(22):8270–75.

465. Deng G, Tang C, Li F, Jiang H, Chen Y. Covalent cross-linked polymer gels with reversible sol–gel transition and self-healing properties. Macromolecules. 2010;43(3):1191–94.

466. Xu Y, Li Y, Chen Q, Fu L, Tao L, Wei Y. Injectable and self-healing chitosan hydrogel based on imine bonds: design and therapeutic applications. International Journal of Molecular Sciences. 2018;19(8):2198.

467. Deng CC, Brooks WL, Abboud KA, Sumerlin BS. Boronic acid-based hydrogels undergo self-healing at neutral and acidic pH. ACS Macro Letters. 2015;4(2):220–24.

468. Liu Y, Liu Y, Wang Q, Han Y, Tan Y. Boronic ester-based self-healing hydrogels formed by using intermolecular BN coordination. Polymer. 2020;202:122624.

469. Canadell J, Goossens H, Klumperman B. Self-healing materials based on disulfide links. Macromolecules. 2011;44(8):2536–41.

470. Mredha MTI, Na JY, Seon J-K, Cui J, Jeon I. Multifunctional poly (disulfide) hydrogels with extremely fast self-healing ability and degradability. Chemical Engineering Journal. 2020;394:124941.

471. Xu K, Lu Y, Takei K. Multifunctional skin-inspired flexible sensor systems for wearable electronics. Advanced Materials Technologies. 2019;4(3):1800628.

472. Pu X, Hu W, Wang ZL. Toward wearable self-charging power systems: the integration of energy-harvesting and storage devices. Small. 2018;14(1):1702817.
473. Liu Z, Mo F, Li H, Zhu M, Wang Z, Liang G, et al. Advances in flexible and wearable energy-storage textiles. Small Methods. 2018;2(11):1800124.
474. Zhang L, Liao M, Bao L, Sun X, Peng H. The functionalization of miniature energy-storage devices. Small Methods. 2017;1(9):1700211.
475. Xu C, Song Y, Han M, Zhang H. Portable and wearable self-powered systems based on emerging energy harvesting technology. Microsystems & Nanoengineering. 2021;7(1):1–14.
476. Pomerantseva E, Bonaccorso F, Feng X, Cui Y, Gogotsi Y. Energy storage: the future enabled by nanomaterials. Science. 2019;366(6468).
477. Shi B, Liu Z, Zheng Q, Meng J, Ouyang H, Zou Y, et al. Body-integrated self-powered system for wearable and implantable applications. ACS Nano. 2019;13(5):6017–24.
478. Leonov V, Vullers RJ. Wearable electronics self-powered by using human body heat: the state of the art and the perspective. Journal of Renewable and Sustainable Energy. 2009;1 (6):062701.
479. Choi Y-M, Lee MG, Jeon Y. Wearable biomechanical energy harvesting technologies. Energies. 2017;10(10):1483.
480. Li C, Jiang F, Liu C, Liu P, Xu J. Present and future thermoelectric materials toward wearable energy harvesting. Applied Materials Today. 2019;15:543–57.
481. Barrade P, Rufer A, editors. Supercapacitors as energy buffers: a solution for elevators and for electric busses supply. Proceedings of the Power Conversion Conference-Osaka 2002 (Cat No 02TH8579); 2002: IEEE.
482. Bi C, Chen B, Wei H, DeLuca S, Huang J. Efficient flexible solar cell based on composition-tailored hybrid perovskite. Advanced Materials. 2017;29(30):1605900.
483. Song K, Han JH, Lim T, Kim N, Shin S, Kim J, et al. Subdermal flexible solar cell arrays for powering medical electronic implants. Advanced Healthcare Materials. 2016;5(13):1572–80.
484. Pagliaro M, Ciriminna R, Palmisano G. Flexible solar cells. ChemSusChem: Chemistry & Sustainability Energy & Materials. 2008;1(11):880–91.
485. Schubert MB, Werner JH. Flexible solar cells for clothing. Materials Today. 2006;9(6):42–50.
486. Park KI, Son JH, Hwang GT, Jeong CK, Ryu J, Koo M, et al. Highly-efficient, flexible piezoelectric PZT thin film nanogenerator on plastic substrates. Advanced Materials. 2014;26 (16):2514–20.
487. Dong Y, Mallineni SSK, Maleski K, Behlow H, Mochalin VN, Rao AM, et al. Metallic MXenes: a new family of materials for flexible triboelectric nanogenerators. Nano Energy. 2018;44:103–10.
488. Hinchet R, Seung W, Kim SW. Recent progress on flexible triboelectric nanogenerators for selfpowered electronics. ChemSusChem. 2015;8(14):2327–44.
489. Fan FR, Luo J, Tang W, Li C, Zhang C, Tian Z, et al. Highly transparent and flexible triboelectric nanogenerators: performance improvements and fundamental mechanisms. Journal of Materials Chemistry A. 2014;2(33):13219–25.
490. Chandrashekar BN, Deng B, Smitha AS, Chen Y, Tan C, Zhang H, et al. Roll-to-roll green transfer of CVD graphene onto plastic for a transparent and flexible triboelectric nanogenerator. Advanced Materials. 2015;27(35):5210–16.
491. Zhong Y, Xia X, Mai W, Tu J, Fan HJ. Integration of energy harvesting and electrochemical storage devices. Advanced Materials Technologies. 2017;2(12):1700182.
492. Kim HS, Kim J-H, Kim J. A review of piezoelectric energy harvesting based on vibration. International Journal of Precision Engineering and Manufacturing. 2011;12(6):1129–41.
493. Du Y, Xu J, Paul B, Eklund P. Flexible thermoelectric materials and devices. Applied Materials Today. 2018;12:366–88.

494. Bahk J-H, Fang H, Yazawa K, Shakouri A. Flexible thermoelectric materials and device optimization for wearable energy harvesting. Journal of Materials Chemistry C. 2015;3 (40):10362–74.

495. Yadav A, Pipe K, Shtein M. Fiber-based flexible thermoelectric power generator. Journal of Power Sources. 2008;175(2):909–13.

496. Chen X, Yin L, Lv J, Gross AJ, Le M, Gutierrez NG, et al. Stretchable and Flexible Buckypaper-Based Lactate Biofuel Cell for Wearable Electronics. Advanced Functional Materials. 2019;29 (46):1905785.

497. Sales FC, Iost RM, Martins MV, Almeida MC, Crespilho FN. An intravenous implantable glucose/dioxygen biofuel cell with modified flexible carbon fiber electrodes. Lab on a Chip. 2013;13(3):468–74.

498. Niiyama A, Murata K, Shigemori Y, Zebda A, Tsujimura S. High-performance enzymatic biofuel cell based on flexible carbon cloth modified with MgO-templated porous carbon. Journal of Power Sources. 2019;427:49–55.

499. Haneda K, Yoshino S, Ofuji T, Miyake T, Nishizawa M. Sheet-shaped biofuel cell constructed from enzyme-modified nanoengineered carbon fabric. Electrochimica Acta. 2012;82:175–78.

500. Nazeeruddin MK, Baranoff E, Grätzel M. Dye-sensitized solar cells: a brief overview. Solar energy. 2011;85(6):1172–78.

501. Hagfeldt A, Boschloo G, Sun L, Kloo L, Pettersson H. Dye-sensitized solar cells. Chemical Reviews. 2010;110(11):6595–663.

502. You J, Chen CC, Dou L, Murase S, Duan HS, Hawks SA, et al. Metal oxide nanoparticles as an electron-transport layer in high-performance and stable inverted polymer solar cells. Advanced Materials. 2012;24(38):5267–72.

503. Kim Y, Ballantyne AM, Nelson J, Bradley DD. Effects of thickness and thermal annealing of the PEDOT: PSS layer on the performance of polymer solar cells. Organic Electronics. 2009;10 (1):205–09.

504. Hu Z, Zhang J, Hao Z, Zhao Y. Influence of doped PEDOT: PSS on the performance of polymer solar cells. Solar Energy Materials and Solar Cells. 2011;95(10):2763–67.

505. Yang HB, Dong YQ, Wang X, Khoo SY, Liu B. Cesium carbonate functionalized graphene quantum dots as stable electron-selective layer for improvement of inverted polymer solar cells. ACS Applied Materials & Interfaces. 2014;6(2):1092–99.

506. Zheng L, Wang J, Xuan Y, Yan M, Yu X, Peng Y, et al. A perovskite/silicon hybrid system with a solar-to-electric power conversion efficiency of 25.5%. Journal of Materials Chemistry A. 2019;7(46):26479–89.

507. Hyun DC, Park M, Park C, Kim B, Xia Y, Hur JH, et al. Ordered zigzag stripes of polymer gel/metal nanoparticle composites for highly stretchable conductive electrodes. Advanced Materials. 2011;23(26):2946–50.

508. Wang C, Zheng W, Yue Z, Too CO, Wallace GG. Buckled, stretchable polypyrrole electrodes for battery applications. Advanced Materials. 2011;23(31):3580–84.

509. Gonçalves C, Ferreira Da Silva A, Gomes J, Simoes R. Wearable e-textile technologies: a review on sensors, actuators and control elements. Inventions. 2018;3(1):14.

510. Lee Y-H, Kim J-S, Noh J, Lee I, Kim HJ, Choi S, et al. Wearable textile battery rechargeable by solar energy. Nano Letters. 2013;13(11):5753–61.

511. Lindström H, Holmberg A, Magnusson E, Malmqvist L, Hagfeldt A. A new method to make dye-sensitized nanocrystalline solar cells at room temperature. Journal of Photochemistry and Photobiology A: Chemistry. 2001; 145 (1–2): 107–12.

512. Zhang D, Yoshida T, Furuta K, Minoura H. Hydrothermal preparation of porous nano-crystalline TiO2 electrodes for flexible solar cells. Journal of Photochemistry and Photobiology A: Chemistry. 2004; 164 (1–3): 159–66.

513. Pan H, Ko SH, Misra N, Grigoropoulos CP. Laser annealed composite titanium dioxide electrodes for dye-sensitized solar cells on glass and plastics. Applied Physics Letters. 2009;94(7):071117.
514. Murakami TN, Kijitori Y, Kawashima N, Miyasaka T. UV light-assisted chemical vapor deposition of TiO2 for efficiency development at dye-sensitized mesoporous layers on plastic film electrodes. Chemistry Letters. 2003;32(11):1076–77.
515. Miyasaka T, Ikegami M, Kijitori Y. Photovoltaic performance of plastic dye-sensitized electrodes prepared by low-temperature binder-free coating of mesoscopic titania. Journal of the Electrochemical Society. 2007;154(5):A455.
516. Yamaguchi T, Tobe N, Matsumoto D, Nagai T, Arakawa H. Highly efficient plastic-substrate dye-sensitized solar cells with validated conversion efficiency of 7.6%. Solar Energy Materials and Solar Cells. 2010;94(5):812–16.
517. Ito S, Rothenberger G, Liska P, Comte P, Zakeeruddin SM, Péchy P, et al. High-efficiency (7.2%) flexible dye-sensitized solar cells with Ti-metal substrate for nanocrystalline-TiO 2 photoanode. Chemical Communications. 2006(38):4004–06.
518. Park JH, Jun Y, Yun H-G, Lee S-Y, Kang MG. Fabrication of an efficient dye-sensitized solar cell with stainless steel substrate. Journal of the Electrochemical Society. 2008;155(7):F145.
519. Chen T, Wang S, Yang Z, Feng Q, Sun X, Li L, et al. Flexible, light-weight, ultrastrong, and semiconductive carbon nanotube fibers for a highly efficient solar cell. Angewandte Chemie. 2011;123(8):1855–59.
520. Lv Z, Yu J, Wu H, Shang J, Wang D, Hou S, et al. Highly efficient and completely flexible fiber-shaped dye-sensitized solar cell based on TiO 2 nanotube array. Nanoscale. 2012;4 (4):1248–53.
521. Wen Z, Yeh M-H, Guo H, Wang J, Zi Y, Xu W, et al. Self-powered textile for wearable electronics by hybridizing fiber-shaped nanogenerators, solar cells, and supercapacitors. Science Advances. 2016;2(10):e1600097.
522. Wu W, Bai S, Yuan M, Qin Y, Wang ZL, Jing T. Lead zirconate titanate nanowire textile nanogenerator for wearable energy-harvesting and self-powered devices. ACS Nano. 2012;6 (7):6231–35.
523. Romano G, Mantini G, Di Carlo A, D'Amico A, Falconi C, Wang ZL. Piezoelectric potential in vertically aligned nanowires for high output nanogenerators. Nanotechnology. 2011;22 (46):465401.
524. Hyeon DY, Park KI. Vertically aligned piezoelectric perovskite nanowire array on flexible conducting substrate for energy harvesting applications. Advanced Materials Technologies. 2019;4(8):1900228.
525. Kumar B, Kim S-W. Energy harvesting based on semiconducting piezoelectric ZnO nanostructures. Nano Energy. 2012;1(3):342–55.
526. Yang R, Qin Y, Dai L, Wang ZL. Power generation with laterally packaged piezoelectric fine wires. Nature Nanotechnology. 2009;4(1):34–39.
527. Hadimani RL, Bayramol DV, Sion N, Shah T, Qian L, Shi S, et al. Continuous production of piezoelectric PVDF fibre for e-textile applications. Smart Materials and Structures. 2013;22 (7):075017.
528. Deterre M, Lefeuvre E, Dufour-Gergam E. An active piezoelectric energy extraction method for pressure energy harvesting. Smart Materials and Structures. 2012;21(8):085004.
529. Rafique S, Rafique Q, Quinn. Piezoelectric Vibration Energy Harvesting: Springer; 2018.
530. Ilyas MA, Swingler J. Piezoelectric energy harvesting from raindrop impacts. Energy. 2015;90:796–806.
531. Choi J, Jung I, Kang C-Y. A brief review of sound energy harvesting. Nano Energy. 2019;56:169–83.

532. Wood R, Steltz E, Fearing R. Optimal energy density piezoelectric bending actuators. Sensors and Actuators A: Physical. 2005;119(2):476–88.

533. Bisegna P, Caruso G. Evaluation of higher-order theories of piezoelectric plates in bending and in stretching. International Journal of Solids and Structures. 2001; 38 (48–49): 8805–30.

534. Delnavaz A, Voix J. Flexible piezoelectric energy harvesting from jaw movements. Smart Materials and Structures. 2014;23(10):105020.

535. Chiu Y-Y, Lin W-Y, Wang H-Y, Huang S-B, Wu M-H. Development of a piezoelectric polyvinylidene fluoride (PVDF) polymer-based sensor patch for simultaneous heartbeat and respiration monitoring. Sensors and Actuators A: Physical. 2013;189:328–34.

536. Liu Z, Zhang S, Jin Y, Ouyang H, Zou Y, Wang X, et al. Flexible piezoelectric nanogenerator in wearable self-powered active sensor for respiration and healthcare monitoring. Semiconductor Science and Technology. 2017;32(6):064004.

537. Wang J, Zhou S, Zhang Z, Yurchenko D. High-performance piezoelectric wind energy harvester with Y-shaped attachments. Energy Conversion and Management. 2019;181:645–52.

538. Lee HJ, Sherrit S, Tosi LP, Walkemeyer P, Colonius T. Piezoelectric energy harvesting in internal fluid flow. Sensors. 2015;15(10):26039–62.

539. Dagdeviren C, Yang BD, Su Y, Tran PL, Joe P, Anderson E, et al. Conformal piezoelectric energy harvesting and storage from motions of the heart, lung, and diaphragm. Proceedings of the National Academy of Sciences. 2014;111(5):1927–32.

540. Fan K, Liu Z, Liu H, Wang L, Zhu Y, Yu B. Scavenging energy from human walking through a shoe-mounted piezoelectric harvester. Applied Physics Letters. 2017;110(14):143902.

541. Xue X, Wang S, Guo W, Zhang Y, Wang ZL. Hybridizing energy conversion and storage in a mechanical-to-electrochemical process for self-charging power cell. Nano Letters. 2012;12 (9):5048–54.

542. Wu C, Wang AC, Ding W, Guo H, Wang ZL. Triboelectric nanogenerator: a foundation of the energy for the new era. Advanced Energy Materials. 2019;9(1):1802906.

543. Kim W-G, Kim D-W, Tcho I-W, Kim J-K, Kim M-S, Choi Y-K. Triboelectric Nanogenerator: structure, Mechanism, and Applications. ACS Nano. 2021;15(1):258–87.

544. Pan S, Zhang Z. Triboelectric effect: a new perspective on electron transfer process. Journal of Applied Physics. 2017;122(14):144302.

545. Pan S, Zhang Z. Fundamental theories and basic principles of triboelectric effect: a review. Friction. 2019;7(1):2–17.

546. Williams MW. Triboelectric charging in metal–polymer contacts–How to distinguish between electron and material transfer mechanisms. Journal of Electrostatics. 2013;71(1):53–54.

547. Xia K, Zhu Z, Zhang H, Du C, Xu Z, Wang R. Painting a high-output triboelectric nanogenerator on paper for harvesting energy from human body motion. Nano Energy. 2018;50:571–80.

548. Yang Y, Zhu G, Zhang H, Chen J, Zhong X, Lin Z-H, et al. Triboelectric nanogenerator for harvesting wind energy and as self-powered wind vector sensor system. ACS Nano. 2013;7 (10):9461–68.

549. Liu Y, Sun N, Liu J, Wen Z, Sun X, Lee S-T, et al. Integrating a silicon solar cell with a triboelectric nanogenerator via a mutual electrode for harvesting energy from sunlight and raindrops. ACS Nano. 2018;12(3):2893–99.

550. Zhang C, Liu L, Zhou L, Yin X, Wei X, Hu Y, et al. Self-Powered Sensor for Quantifying Ocean Surface Water Waves Based on Triboelectric Nanogenerator. ACS Nano. 2020;14(6):7092–100.

551. Pu X, Li L, Song H, Du C, Zhao Z, Jiang C, et al. A self-charging power unit by integration of a textile triboelectric nanogenerator and a flexible lithium-ion battery for wearable electronics. Advanced Materials. 2015;27(15):2472–78.

552. Gao T, Zhao K, Liu X, Yang Y. Implanting a solid Li-ion battery into a triboelectric nanogenerator for simultaneously scavenging and storing wind energy. Nano Energy. 2017;41:210–16.
553. Pei Y, LaLonde A, Iwanaga S, Snyder GJ. High thermoelectric figure of merit in heavy hole dominated PbTe. Energy & Environmental Science. 2011;4(6):2085–89.
554. Sabarinathan M, Omprakash M, Harish S, Navaneethan M, Archana J, Ponnusamy S, et al. Enhancement of power factor by energy filtering effect in hierarchical BiSbTe3 nanostructures for thermoelectric applications. Applied Surface Science. 2017;418:246–51.
555. Toprak MS, Stiewe C, Platzek D, Williams S, Bertini L, Müller E, et al. The impact of nanostructuring on the thermal conductivity of thermoelectric CoSb3. Advanced Functional Materials. 2004;14(12):1189–96.
556. Kambe K, Udono H. Convenient Melt-Growth Method for Thermoelectric Mg 2 Si. Journal of Electronic Materials. 2014;43(6):2212–17.
557. Lee H, Vashaee D, Wang D, Dresselhaus MS, Ren Z, Chen G. Effects of nanoscale porosity on thermoelectric properties of SiGe. Journal of Applied Physics. 2010;107(9):094308.
558. You H, Jin Y, Chen J, Li C. Direct coating of a DKGM hydrogel on glass fabric for multifunctional oil-water separation in harsh environments. Chemical Engineering Journal. 2018;334:2273–82.
559. Kim SJ, We JH, Cho BJ. A wearable thermoelectric generator fabricated on a glass fabric. Energy & Environmental Science. 2014;7(6):1959–65.
560. Lu Z, Zhang H, Mao C, Li CM. Silk fabric-based wearable thermoelectric generator for energy harvesting from the human body. Applied energy. 2016;164:57–63.
561. Deng F, Qiu H, Chen J, Wang L, Wang B. Wearable thermoelectric power generators combined with flexible supercapacitor for low-power human diagnosis devices. IEEE Transactions on industrial electronics. 2016;64(2):1477–85.
562. Lee J-H, Kim J, Kim TY, Al Hossain MS, Kim S-W, Kim JH. All-in-one energy harvesting and storage devices. Journal of Materials Chemistry A. 2016;4(21):7983–99.
563. Lingam D, Parikh AR, Huang J, Jain A, Minary-Jolandan M. Nano/microscale pyroelectric energy harvesting: challenges and opportunities. International Journal of Smart and Nano Materials. 2013;4(4):229–45.
564. Bhalla A, Fang C, Xi Y, Cross L. Pyroelectric properties of the alanine and arsenic-doped triglycine sulfate single crystals. Applied Physics Letters. 1983;43(10):932–34.
565. Levstik A, Golob B, Kosec M. Dielectric and pyroelectric properties of Pb5Ge3O11 ceramics. Journal of Applied Physics. 1992;71(8):3922–25.
566. Gesualdi M, Jacinto C, Catunda T, Muramatsu M, Pilla V. Thermal lens spectrometry in pyroelectric lithium niobate crystals. Applied Physics B. 2008;93(4):879–83.
567. Ma J, Wu Z, Luo W, Zheng Y, Jia Y, Wang L, et al. High pyrocatalytic properties of pyroelectric BaTiO3 nanofibers loaded by noble metal under room-temperature thermal cycling. Ceramics International. 2018;44(17):21835–41.
568. Kao M-C, Lee M-S, Wang C-M, Chen H-Z, Chen Y-C. Properties of LiTaO3 thin films derived by a diol-based sol–gel process. Japanese Journal of Applied Physics. 2002;41(5R):2982.
569. Mischenko A, Zhang Q, Scott J, Whatmore R, Mathur N. Giant electrocaloric effect in thin-film PbZrO. 95TiO. 05O3. Science. 2006;311(5765):1270–71.
570. Shaldin YV, Matyiasik S, Davydov A, Zhavoronkov N. Pyroelectric properties of the wide-gap CdSe semiconductor in the low-temperature region. Semiconductors. 2014;48(1):1–8.
571. Liu J, Fernández-Serra MV, Allen PB. First-principles study of pyroelectricity in GaN and ZnO. Physical Review B. 2016;93(8):081205.
572. Zhang M, Hu Q, Ma K, Ding Y, Li C. Pyroelectric effect in CdS nanorods decorated with a molecular Co-catalyst for hydrogen evolution. Nano Energy. 2020;73:104810.

573. Xue H, Yang Q, Wang D, Luo W, Wang W, Lin M, et al. A wearable pyroelectric nanogenerator and self-powered breathing sensor. Nano Energy. 2017;38:147–54.
574. Lee JH, Lee KY, Gupta MK, Kim TY, Lee DY, Oh J, et al. Highly stretchable piezoelectric-pyroelectric hybrid nanogenerator. Advanced Materials. 2014;26(5):765–69.
575. Zi Y, Lin L, Wang J, Wang S, Chen J, Fan X, et al. Triboelectric–pyroelectric–piezoelectric hybrid cell for high-efficiency energy-harvesting and self-powered sensing. Advanced Materials. 2015;27(14):2340–47.
576. Vaish M, Sharma M, Vaish R, Chauhan VS. Capacitor and battery charging from hot/cold air using pyroelectric ceramics. Integrated Ferroelectrics. 2016;176(1):160–70.
577. Bandodkar AJ, Wang J. Wearable biofuel cells: a review. Electroanalysis. 2016;28(6):1188–200.
578. Bandodkar AJ. Wearable biofuel cells: past, present and future. Journal of the Electrochemical Society. 2016;164(3):H3007.
579. Choi Y, Wang G, Nayfeh MH, Yau S-T. A hybrid biofuel cell based on electrooxidation of glucose using ultra-small silicon nanoparticles. Biosensors and Bioelectronics. 2009;24(10):3103–07.
580. Lv J, Jeerapan I, Tehrani F, Yin L, Silva-Lopez CA, Jang J-H, et al. Sweat-based wearable energy harvesting-storage hybrid textile devices. Energy & Environmental Science. 2018;11(12):3431–42.
581. Bandodkar A, Lee S, Huang I, Li W, Wang S, Su C-J, et al. Sweat-activated biocompatible batteries for epidermal electronic and microfluidic systems. Nature Electronics. 2020;3(9):554–62.

Index

3D printing 39

amorphous phase 68
amplitude 15
anisotropic crystal 94
aqueous electrolytes 46
array drawing 47

bending angle 11
bending energy 13
biocompatible 3–4, 86, 97
block copolymer 4
boiling point 65
breading 86

capacitance retention 22, 27, 29, 35, 77
carbon aerogels 22
carbon nanotube 22
ceramic electrolyte 55
charge transport kinetics 48
charge/discharge reaction 50
chemical bonding 26
CO_2 reduction 53
coaxial structure 24
coiled interconnect 9
Collapsing radius geometry 11
compression molding 4
compressive strain 10
condensation 80
conductive filler 6, 46
conjugated polymer 85
contact electrification 90
contact resistance 5
co-ordination bond 78
co-ordination number 78
copolymer 4
core sheath fiber 28
corrosion 52
corrosion inhibitor 52
Coulombic efficiency 45
covalent organic frameworks 23
crown ether 79
crystal lattice 89
crystal structure 88
Curie temperature 89
current collector 2

cycle life 22
cycling stability 53
cystic fibrosis 1
cytotoxicity 96

DC–DC boost converter 98
deintercalation 46
diagnosis 89
dielectric constant 65
discharge capacity 53
Doping 57
drug delivery 100
ductility 19

elastic limit 10
elastic modulus 4, 10
elasticity 10
electrical conductivity 2–3, 5, 22, 26–29, 32–33,
    37, 48, 51, 57, 60, 62, 73, 81, 86, 92
electrical dipole 89
electrical double-layer capacitance 26
electrocatalyst 50
electrochemical reaction 3, 53
electrodeposition 26
electroless coating 26
electromagnetic induction 98
electromechanical interaction 88
electron-beam irradiation technique 72
electronic skin 57
electrosorption 22
electrospinning 23, 26, 30, 52, 54–55, 73
electrospraying 23
electrostatic induction 90
embroidering 86
emulsion 5
energy density 2, 22, 24, 27–28, 32, 34, 37,
    44–45, 48, 50–51, 53–56, 60, 64, 69, 77,
    83, 96
e-textile technologies 86
Euler buckling 18
extrusion molding 4

ferroelectric 89
figure of merit 92
film patternability 5
film thickness 5

https://doi.org/10.1515/9781501521287-007

www.ingramcontent.com/pod-product-compliance
Lightning Source LLC
Chambersburg PA
CBHW081542220326
41598CB00036B/6526